Lecture Notes in Business Information Processing 296

Series Editors

Wil M.P. van der Aalst
Eindhoven Technical University, Eindhoven, The Netherlands
John Mylopoulos
University of Trento, Trento, Italy
Michael Rosemann
Queensland University of Technology, Brisbane, QLD, Australia
Michael J. Shaw
University of Illinois, Urbana-Champaign, IL, USA
Clemens Szyperski
Microsoft Research, Redmond, WA, USA

Ming Fan · Jukka Heikkilä
Hongxiu Li · Michael J. Shaw
Han Zhang (Eds.)

Internetworked World

15th Workshop on e-Business, WeB 2016
Dublin, Ireland, December 10, 2016
Revised Selected Papers

 Springer

Editors
Ming Fan
University of Washington
Seattle, WA
USA

Michael J. Shaw
University of Illinois
Urbana-Champaign, IL
USA

Jukka Heikkilä
University of Turku
Turku
Finland

Han Zhang
Georgia Institute of Technology
Atlanta, GA
USA

Hongxiu Li
University of Turku
Turku
Finland

ISSN 1865-1348   ISSN 1865-1356   (electronic)
Lecture Notes in Business Information Processing
ISBN 978-3-319-69643-0   ISBN 978-3-319-69644-7   (eBook)
https://doi.org/10.1007/978-3-319-69644-7

Library of Congress Control Number: 2017957560

This Springer imprint is published by Springer Nature
The registered company is Springer International Publishing AG
The registered company address is: Gewerbestrasse 11, 6330 Cham, Switzerland

Preface

The Workshop on e-Business (WeB) is a premier annual workshop on e-business and e-commerce. The purpose of the workshop is to provide an open forum for e-business researchers and practitioners worldwide to explore and respond to the challenges of next-generation e-business systems, share the latest research findings, explore novel ideas, discuss success stories and lessons learned, map out major challenges, and collectively chart the future directions of e-business. Since its inception in 2000, the WeB workshop has attracted state-of-the-art research and followed closely the developments in the technical and managerial aspects of e-business. The 15th Annual Workshop on e-Business (WeB 2016) was held in Dublin, Ireland, on December 10, 2016.

The theme of WeB 2016 was "Internetworked World." Digitalization, consumerization, global platforms, and transformative innovations are causing industry convergence at a record pace. The lifecycle of companies has shifted, restructuring can happen overnight, and in the leading cases there are enormous benefits to be reaped – if the partnering e-business and social media networks/platforms are constructed properly and effectively.

This has opened up wholly new opportunities for consumers to take advantage of global supply of goods and services over the Internet and logistics, resulting in the growth of a new generation of Internet consumers/businesses. Simultaneously, innovative nations are establishing cooperation between industry, government, and consumers/citizens to boost service co-creation and to transform public and private sector efficiency, thus creating new jobs, services, and businesses. The novel technologies are making it feasible for both new entrants and incumbents to try innovative ideas within their businesses with the help of global platforms and networks. The consumer responses, business transactions, and market sentiments can be sensed in real time in this internetworked world and tracked via data analytics. WeB 2016 provided a forum for scholars to exchange ideas and share results from their research on the aforementioned theme. Original research articles addressing a broad coverage of technical, managerial, economic, and strategic issues related to consumers, businesses, industries, and governments were presented at the workshop. These articles employed various IS research methods such as case study, survey, analytical modeling, experiments, computational models, design science, etc.

We received 46 submissions and each submission was reviewed by three reviewers. The Program Committee co-chairs had a final consultation meeting to look at all the reviews and make the final decisions on the papers to be accepted. We accepted 23 papers (50%), including 15 papers as long/regular papers and 8 as short papers.

We would like to thank all the reviewers for their time and effort and for completing their review assignments on time despite tight deadlines. Many thanks to the authors for their contributions.

September 2017

Ming Fan
William Golden
Jukka Heikkilä
Hongxiu Li
Michael J. Shaw
Han Zhang

Organization

Honorary Chair

Andrew B. Whinston   University of Texas at Austin, USA

Conference Chairs

William Golden   NUI Galway, Ireland
Jukka Heikkilä   University of Turku, Finland
Michael J. Shaw   University of Illinois at Urbana-Champaign, USA

Program Co-chairs

Ming Fan   University of Washington, USA
Hongxiu Li   University of Turku, Finland
Han Zhang   Georgia Institute of Technology, USA

Local Arrangements Chair

Hongxiu Li   University of Turku, Finland

Program Committee

Subhajyoti Bandyopadhyay	University of Florida, USA
Joseph Barjis	Delft University of Technology, The Netherlands
Hsinlu Chang	National Chengchi University, Taiwan
Michael Chau	The University of Hong Kong, SAR China
Patrick Y.K. Chau	The University of Hong Kong, SAR China
Kenny Cheng	University of Florida, USA
Ching-Chin Chern	National Taiwan University, Taiwan
Honghui Deng	University of Nevada, Las Vegas, USA
Aidan Duane	Waterford Institute of Technology, Ireland
Ming Fan	University of Washington, USA
Xianjun Geng	University of Texas, Dallas, USA
Zhiling Guo	Singapore Management University, Singapore
Wencui Han	University of Illinois at Urbana-Champaign, USA
Lin Hao	University of Notre Dame, USA
Yuheng Hu	University of Illinois at Chicago, USA
Jinghua Huang	Tsinghua University, China
Prasanna Karhade	Hong Kong University of Science and Technology, SAR China

Dan Ke	Wuhan University, China
Jan Kraemer	Karlsruhe Institute of Technology, Germany
Karl Reiner Lang	The City University of New York, Baruch College, USA
Eric Larson	University of Illinois at Urbana-Champaign, USA
Anthony Lee	National Taiwan University, Taiwan
Hongxiu Li	University of Turku, Finland
Xitong Li	HEC, Paris, France
Zhangxi Lin	Texas Tech University, USA
Nirup Menon	George Mason University, USA
Matt Nelson	Illinois State University, USA
Dirk Neumann	University of Freiburg, Germany
Chih-Hung Peng	City University of Hong Kong, SAR China
Selwyn Piramuthu	University of Florida, USA
Liangfei Qiu	University of Florida, USA
Jackie Rees	Iowa State University, USA
Michael Rosemann	Queensland University of Technology, Australia
Huaxia Rui	University of Rochester, USA
Raghu T. Santanam	Arizona State University, USA
Ravi Sen	Texas A&M University, USA
Benjamin Shao	Arizona State University, USA
Riyaz Sikora	University of Texas at Arlington, USA
Chandrasekar Subramaniam	University of North Carolina Charlotte, USA
Ram Subramanyam	University of Illinois at Urbana-Champaign, USA
Vijayan Sugumaran	Oakland University, USA
Kar Yan Tam	Hong Kong University of Science and Technology, SAR China
Yinliang Tan	Tulane University, USA
James Thong	Hong Kong University of Science and Technology, SAR China
Kai Wang	National University of Kaohsiung, Taiwan
Chih-Ping Wei	National Taiwan University, Taiwan
Lizhen Xu	Georgia Institute of Technology, USA
Dezhi Yin	University of Missouri, USA
Byungjoon Yoo	Seoul National University, South Korea
Victoria Yoon	Virginia Commonwealth University, USA
Han Zhang	Georgia Institute of Technology, USA
Michael Zhang	Hong Kong University of Science and Technology, SAR China
Kexin Zhao	University of North Carolina Charlotte, USA
Wei Zhou	ESCP Europe
Dan Zhu	Iowa State University, USA

Contents

A Consumer-Oriented Decision-Making Approach for Selecting the Cloud Storage Service: From PAPRIKA Perspective

Salim Alismaili[1(✉)], Mengxiang Li[2], Jun Shen[1], and Qiang He[2]

[1] School of Computing and Information Technology, University of Wollongong,
Wollongong, Australia
szaai787@uowmail.edu.au, jshen@uow.edu.au
[2] School of Software and Electrical Engineering,
Swinburne University of Technology, Melbourne, Australia
mli@uow.edu.au, qhe@swin.edu.au

Abstract. In recent years there is a growth in the number of companies that offers cloud storage solutions. From user's perspectives, it is becoming a challenging task to choose which cloud storage to use and from whom, based on user's needs. In this context, no framework can evaluate the decision criteria for selection of cloud storage services. This paper proposes a solution to this problem by identifying the cloud storage criteria and introduces the PAPRIKA approach for measuring the criteria of cloud storage based on client's preference. This work demonstrated the applicability of the framework (decision model) by testing it with eleven users of cloud storage services. The results showed that the model could help users in making a more informative decision about cloud storage services.

Keywords: Cloud storage · Decision making · PAPRIKA approach

1 Introduction

Cloud computing is a contemporary computing concept for conveying on-demand resources (e.g., infrastructure, platform, and software) to customers. The cloud computing services include: Software as a Service (SaaS), where customer access to applications that run on providers infrastructure; Platform as a Service (PaaS) where customer use provider's resources to develop applications or run custom applications; Infrastructure as a Service (IaaS) where customer use provider's environment provides services such as storage and networking infrastructure [1]. Our focus is the cloud storage which is a part of IaaS. Services providers usually offer IaaS storage with scalability option either up or down based on user's demand. Cloud storage permits users to store their data to an online server and access them remotely from anywhere. Data security and availability are some of the concerns from the user's perspective for the information warehoused in the cloud; some cloud storage providers solved these anxieties. For instance, Google Drive implemented a two-stage verification for ensuring additional security measure [2].

© Springer International Publishing AG 2017
M. Fan et al. (Eds.): WeB 2016, LNBIP 296, pp. 1–12, 2017.
https://doi.org/10.1007/978-3-319-69644-7_1

An informative decision making in the selection of the best cloud storage is crucial for any user even when the technologies offered for free. In some situations, it is even better in deciding on buying a service due to the supplementary features such as collaboration and higher security protections.

With the broad availability of cloud storage options in the market, it is becoming difficult for users to decide on the right option for them, even if the options are free of cost. Several parameters need to be considered such as cost, storage space, support, security, and reliability. Theses parameters in this context can be either quantitative such as cost and storage capacity or qualitative such as reliability and support. Currently, some users might decide on a certain cloud service based on other users review, other users might decide based on cost, another user may build his decision based on storage capacity. There is no wider framework which considers various parameters and rates them based on user preferences either individually or collectively for a group decision. In a global survey conducted in 2014 and involved 26,000 consumers indicated the importance of cloud storage to the majority of the participants irrespective of their gender group [3]. This implies the importance of cloud storage and the importance of this consumer segment. A forecast of personal cloud storage consumers worldwide estimated to reach to US $ 1.8 billion people in 2017 [4] and it will continue in growing in the coming years. These data provide an indication of the size of this market and, therefore, the importance of this study. Despite the increased usage of cloud storage, there is no framework for assessing different cloud storages available in the market based on individual users' requirements. Therefore, we state our problem as **it is difficult for consumers to make an informative decision on the appropriate cloud storage for their needs.**

Quality measures assist in identifying which of the available cloud storage is the best and meets users' needs. Because of their significance, we selected the certain criteria that are based on ISO/IEC 94126 and a review of 25 websites of cloud services providers. ISO usability models do not cover all usability features [5]. Therefore, websites review was a necessary phase of this study. Some of the criteria were common and existed in both the ISO/IEC 94126 and from our review of the websites. The other criteria which will be evaluated and used to design our cloud storage decision model are storage space, support, upload and download speed, security measures, cost, and compatibility with different devices. Our decision model intends to help in providing a mechanism for ranking these criteria based on either individual or group preferences. The model is capable of generating each single user preference values. However, this paper presents only the results achieved by the panel (the case: University of Wollongong students) on their collective decision about their preference on the mentioned criteria.

2   Related Work

2.1   Cloud Storage

Allows users to store their data online and access them over the network from anywhere and through various computing interfaces. The hosting organisation which is the cloud storage provider installs a client application on a user's device. This application transfers any files that the users desires via the internet to the service provider then the file can be accessed and shared to other user's devices seamlessly and conveniently. Data can be synchronised to an auto update on all systems where the storage software is installed. Cloud storage have significant advantages in providing data accessibility, the reliability of services in comparing to the traditional storage solutions, secure storage, and disaster recovery. Also, the cloud storage cost is lower when compared with having the traditional storage setup with expensive hardware which enquiries management and maintained overheads. Cloud storage can be used to do various operations such as sharing files without the need to send email attachments, storing and accessing of data, additional backup and virtual collaboration with other users. Cloud storage providers are competing in delivering robust techniques for backing up and archiving of data in a secure, reliable, and practical environment. With the promised benefits, cloud storage has brought new concerns which were not there in the past computing environment such as security concerns and compliance issues. As a system cloud storage consists of three things: applications, platform, and the infrastructure. Cloud storage is offered in the three known types cloud computing: public, private, and hybrid. It is also delivered to individual users (personal). In general, cloud storage has several features such as resource pooling and multi-tenancy, scalability of the storage, Operational expenses (OPEX) costing model, sharing files, and collaboration [6]. Data transmitted to cloud storage in two ways either through web-based applications or web services application programming interfaces (APIs). Web-based applications are mainly applied for manual access to data, while APIs are used for managing automated processes [6]. Our focus in this paper is the evaluation of the personal cloud storage solutions.

2.2   Cloud Storage Ranking

Comparing and selection of cloud storage services have not been an easy task due to the wide availability of these services in different forms and with various suppliers. Sometimes the decision of the users is based on the volume of free storage space while other users might be more interested in the service that offers high security. It is believed that there are various factors can influence the embracing or use of cloud services might look in. In the next section, the paper will present seven factors which are found to be the most influential factors to users based on our tracking of user's reviews about the cloud storage services.

A study by Walker, Brisken [7] investigated investment option to lease cloud storage or to buy hard disk drives. Walker and colleagues presented a new modelling method based on comparing purchase cost versus leasing cost of cloud services and used empirical data. Ruiz-Alvarez and Humphrey [8] presented a mathematical model for allocation of datasets in cloud computing. Their model focused on the cost, performance, and the characteristics of cloud computing. Garg, Versteeg [9] used Analytical Hieratical Process (AHP) to rank cloud services based on customer's quality of services (QoS) requirements regarding criteria such as security and performance. According to our best knowledge, those are the few studies which could be relevant to this research; however, each study addressed a specific theme. It is believed that this work is the first attempt at providing an approach for ranking of cloud storage services for individual users. This work addresses cloud storages decision modelling and provides a decision approach based on PAPRIKA methodology taking into account individual user's requirements. This topic has not yet covered in the literature.

2.3   Selection Criteria for Cloud Storage

This work is a first attempt in this direction. The foundation of selecting the criteria (parameters) for this research is based on two aspects (a) International Standard Organisation ISO/IEC 9126 quality guidelines comprises a set of business based key performance indicators that are useful in evaluating services such as cloud storage, and (b) a review of twenty-five most popular cloud storage providers such as JustCloud, Zip Cloud, Google Drive, Microsoft OneDrive, IDrive, and Dropbox. The standard defined five characteristics *(i)* functionality, which explains the presence of multifunction's and their features; *(ii)* reliability, states the capability of software to sustain its performance level under defined conditions and time frame; *(iii)* usability, which is the extent of usage determination; *(iv)* efficiency, which bears the association of the applied resources with the level of software performance; *(v)* maintainability, which endures on the amount of effort consumed to make the intended alteration to the software; and *(vi)* portability, which is the degree of the transferability of the software from one platform to another [10].

There are still no definitions or measures of the identified attributes. The followings define these criteria:

Security: The degree of efficiency and protection of cloud storage services regarding access control, data privacy and confidentially. CSMIC [11] included various attributes that fall under this category including access control, physical and environmental security, and security management. Ensuring the security of transmission channels, methods, and the physical storage location is essential. Security requires using encryption, authorisation, and authentication measures [6].

Reliability: The capability of cloud storage provider to maintain its performance level without failure during a certain time and stated conditions. Cloud services providers should be able to provide a sufficient amount of assurance and demonstrate the acceptable extent of stability and resilience of their services to their clients.

Support: Refer to the technical assistance provided by the cloud storage providers to its customers. Support can be provided using different means of communication.

Storage space: The amount of space available for storing data and it is measured in gigabytes.

Cost: Refers to the cost of cloud storage as per the specification offered. This might be the first concern appears in users' minds before they decide to adopt cloud storage. Using cloud storage could leverage savings on purchasing the traditional storage devices depending on the utilisation scope of each individual. Cloud storages offered at different plans and prices.

Speed (uploads and downloads): Refers to the response time. It is the unit of time takes to upload and download a unit of data. There is a variation of upload and download speed of the online storage services in the market.

Ease of use: Refers to the smoothness of using the cloud storage services in term of aspects such as data management, uploading and downloading files and folders, and accessibility. This criterion plays a vital role in the diffusion rate of cloud services. According to CSMIC [11], many factors such as accessibility, learnability, installability, operability, and transparency can fall in this category.

3   Cloud Storage Decision Modelling

The identified criteria and their specified levels based on the evaluation of the cloud storage providers and user's reviews and comments are mapped in Table 1 to design the decision model for this research.

The criteria level rankings start with lowest ranked to highest ranked as illustrated in Table 1. For example, for the cost criteria, the highest rank is identified to be affordable, and the lowest rank is when the cost of storage determined to be expensive. This paper aims to provide users with a framework to improve the decision-making process with more knowledgeable insights. The model will be tested with eleven cases to validate its functionality and applicability.

Table 1. Cloud storage decision model

Criteria	Rank	Level	
Storage space	Lowest ranked	Sufficient	Cloud
	Highest ranked	High	Storage
Upload and download speed	Lowest ranked	Reasonable	service
	Highest ranked	Fast	
Compatibility with PC, MAC, and mobile devices	Lowest ranked	Only Compatible with either PC or MAC	
	Highest ranked	Fully compatible with PC, MAC, and mobile devices	
Reliability	Lowest ranked	Reasonable (95%)	
	Highest ranked	High (99%)	
Security measures	Lowest ranked	Reasonable secured (98% secured)	
	Highest ranked	Highly secured (99.99% secured)	
Ease of use	Lowest ranked	Requires little knowledge	
	Highest ranked	Easy to use	
Support	Lowest ranked	Phone and email	
	Highest ranked	Phone, live chat, video tutorials, and email	
Cost	Lowest ranked	Expensive	
	Highest ranked	Affordable	

4 Methodology

This paper proposing a ranking of cloud storage services using "Potentially All Pair-wise RanKings of all possible Alternatives" (PAPRIKA) approach. The approach is the foundation for designing and developing a Multi-Criteria Decision Modelling (MCDM) for the cloud storage services ranking. Designing the model requires identifying the relevant criteria that have an impact on the usage of cloud storage services. Those criteria have been developed through examination of the user's reviews of the selection parameters available on twenty-five cloud storage providers. The criteria and their identified levels of preferences are the building blocks of the decision model for this research. The decision model was structured to reveal the relative importance (weights) of the criteria. This was achieved by the input of the participants revealing their preference on the criteria by responding to several questions which involved trade-offs between the criteria at the decisions step.

The evaluation of the preferences in the PAPRIKA method was achieved through the trade-off between all the criteria. The participants had three options to choose between every two compared criteria. These options are "pair one is better than pair two," "pair two is better than pair one," and "both pairs are equal" (Fig. 1).

Fig. 1. Example of a pair-wise ranking trade-off question

The evaluated criteria for this study are qualitative in nature. The relative importance of each criterion is determined by the highest ranked of its preference level, and the total of all the highest preference levels are equal to 100% (data available upon request).

PAPRIKA method is closer to human natural daily decision process as it is associated with comparing between two alternatives at a time. In this sense, it is more vigorous than AHP because AHP is based on 1–9 scaling system. PAPRIKA can theoretically rank any number of alternatives. By this way, PAPRIKA provides a more comfort in the final achieved decision. This research followed the following steps in modelling the cloud storage decision process:

1. Activity model: preparation of the model setup was established.
2. The activity design: for revealing the relative importance of the criteria.
3. Defining criteria: the criteria and their categories (levels) were implemented. The levels were also ranked based on the author's intuition that relies mainly on the review of the cloud storage providers and common sense.
4. Decisions: trading-off between criteria to reveal the preference value for each criterion for each survey participant.
5. Preference values and criterion ranking: the weight of importance of the criteria as determined by the participant as induvial or in a group.

PAPRIKA methodology was proposed to develop and test the cloud storage decision model for several considerations. First of all, its platform provides an easy to develop and deploy the decision model. Second, it reflects the natural human decision on a comparison between only two criteria at a time not like other methodology which evaluates several parameters and alternatives at once. Third, the survey development is cost efficient and clear. Fourth, the structure of the question is direct and efficient. Fifth, the approach is useful for subjective topics similar to this case as different people have different opinion and preferences in the features and type of cloud storage. Sixth, the method entails ranking of opposing substitutes through assessing all potential undominated pairs of criteria, arriving at more concrete results with a useful model [12]. Seventh, the method handles only two criteria to select among at a time, while SMART/SWING (Simple Multi-Attribute Rating Technique using Swing weights), outranking, and some CA (Conjoint Analysis) techniques use collective computations of the criteria to rank alternatives. This makes PAPRIKA resembling human intellect in making decisions because it is naturally easier to determine on a choice when there are fewer options for selection. In this sense, Forman and Selly [13] mentioned that accuracy in the scoring of alternatives depends on decision makers perception and their conception of the scoring scale. Eighth, PAPRIKA can incubate wider preference options than the majority of other alternative scoring methods [12], such as Discrete Choice Experiments/Conjoint Analysis (DCE/CA), Adaptive Conjoint Analysis (ACA), and the Analytical Hierarchy Process (AHP) [14]. For this paper, we propose PAPRIKA methodology; we argue that this method is proper for modelling the cloud storage.

4.1   Survey

The preference survey which was linked with the decision model was then distributed to the participants to reveal their collective decision on the ranking of the criteria. The survey was conducted online using PAPRIKA method through its interface named 1000Minds software [15]. Participants have the option to resume from their stopping point in the survey whenever they are ready. The update is occurring automatically with every newly completed survey for instant analysis. As this methodology provides the trade-off between only two criteria at a time, it, therefore, reduces the issues of participant's bias in answering the survey question without a careful reading of the questions. This bias issue is a common problem with other forms of questionnaire due to participants fatigue related to complexity and length of the surveys [16].

4.2   Participants

The participants of this study are students at the University of Wollongong/Australia who already used cloud storage. Their contact details were obtained randomly through a direct approach to various students in the campus. The students were from different disciplines. Data collected in July 2016. Following Macefield [17] guidelines in participates size, the responses from the eleven cases obtained in this study are sufficient to achieve the objective of testing the usefulness and applicability of the method (see Table 2).

Table 2.   Participants progress report

Progress	Participants
Excluded from activity	0
Email not sent yet (or no address)	2
Email sent, not started yet	29
Started (not finished yet)	6
Finished	11

4.3   Cloud Storage Choice Modelling

The cloud storage model was the foundation for running the preference surveys or the discrete choice experiments. The distribution activity of the survey was done through the model itself using the 1000minds software platform. Within this model, two actions have been carried out: (i) discovering participants weight values of the criteria (ii) ranking the criteria.

5   Results and Discussion

The paper presents the results of the criterion rankings for the eleven participants who completed 100% of the preference survey (preference values available with the authors upon request).

5.1   Criterion Rankings

Table 3 demonstrates the ranking of the criteria for each of the participants with the group achieved median and mean values. It is understandable that every participant had a different opinion in their preference on ranking of the criteria due to the subjectivity of the topic.

Table 3. Criterion rankings

	Participants											Median	Mean
	153575	153573	153574	153544	153551	153553	153555	153556	153550	153547	153546		
Storage space	3rd	1st=	4th	2nd=	2nd	4th	1st	2nd	5th	1st	7th	2.5	3.045
Ease of use	5th	5th	6th	6th=	4th	2nd	3rd	3rd	1st=	3rd	1st	3	3.682
Support	4th	6th	1st	2nd=	1st	7th	4th=	5th	4th	6th	4th	4	4.136
Upload/download speed	7th	7th=	3rd	6th=	5th	1st	4th=	1st	6th=	4th	2nd	5	4.455
Security measures	1st	1st=	2nd	1st	7th	8th	4th=	7th	6th=	8th	3rd	5	4.591
Cost	8th	1st=	5th	4th=	6th	6th	2nd	8th	1st=	2nd	6th	5	4.636
Reliability	2nd	4th	7th	4th=	8th	3rd	7th	6th	8th	5th	8th	6	5.682
Compatibility with PC, MAC, mobile devices	6th	7th=	8th	6th=	3rd	5th	8th	4th	3rd	7th	5th	6	5.773

5.2 Criterion Weights

Figure 2 presents the criteria weights and the mean value (i.e., the thicker black line). It is evident that the storage space is the most important criteria with a weight value of 16.4% and compatibility is the least valuable criteria with a weight value of 9.1%.

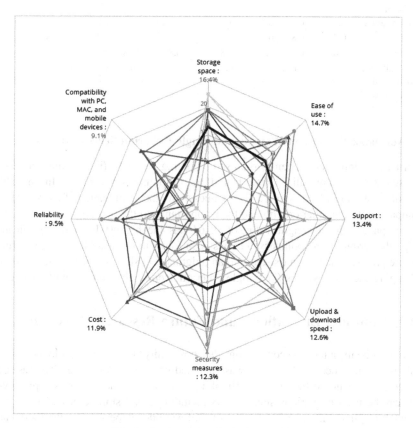

Fig. 2. Radar chart of criterion weights

5.3   Relative Importance of Criteria (Mean Weights)

Table 4 presents additional visualisation output to present an easy to read tabulation for users and decision makers for more informative decision. The table considers the marginal rate of substitution (ratio) of the column criterion for the row criterion. For example, (row 1, col2: 1.1) indicates that 'storage space' was more significant to participants for 1.1 than the 'ease of use' issues and (row 8, col6: 0.8) shows that 'compatibility' forms 0.8 of importance to the 'cost.'

Table 4. Relative importance of the criteria (mean weights)

	Storage space	Ease of use	Support	Upload and download speed	Security measures	Cost	Reliability	Compatibility with PC, MAC, and mobile devices
Storage space		1.1	1.2	1.3	1.3	1.4	1.7	1.8
Ease of use	0.9		1.1	1.2	1.2	1.2	1.5	1.6
Support	0.8	0.9		1.1	1.1	1.1	1.4	1.5
Upload & download speed	0.8	0.9	0.9		1	1.1	1.3	1.4
Security measures	0.8	0.8	0.9	1		1	1.3	1.4
Cost	0.7	0.8	0.9	0.9	1		1.3	1.3
Reliability	0.6	0.6	0.7	0.8	0.8	0.8		1
Compatibility with PC, MAC, and mobile devices	0.6	0.6	0.7	0.7	0.7	0.8	1	

5.4   Normalised Criterion Weights and Single Criterion Scores (Means)

The criteria weights have been normalised summing to 100% (i.e. 1) and single criterion scores with a normalised scale from 0 to 100 points (see Table 5). In this table, the points system has been applied. The values show the significance of each criterion in comparison to the other criteria and their importance to the participants. Apparently, 'storage space' with a weight value of 0.164 has the highest level of significance among other criteria. The points system has been found to be easier to apply, vigorous, and more precise than the unassisted human judgments [18]. It is worth to mention that by changing the point values (i.e. single criterion scores), the ranking will change.

6   Contribution, Limitations, and Future Research Direction

This paper contributed in confirming the possibility to model a cloud storage decision-making process. The model was validated with real world cases. The research involved participants in the process of ranking criteria. The model forms a prototype which can be used by various stakeholders including cloud storage providers to gain additional insights to improve their services. Software developers to use the model and

Table 5. Normalised criterion weights and single criterion scores (means)

Criterion	Criterion weight	Level	Single criterion
	(sum to 1)		Score (0–100)
Storage space	0.164	Sufficient	0
		High	100
Upload and download speed	0.126	Reasonable	0
		Fast	100
Compatibility with PC, MAC, and mobile devices	0.091	Only Compatible with either PC or MAC	0
		Fully compatible with PC, MAC, and mobile devices	100
Reliability	0.095	Reasonable (95%)	0
		High (99%)	100
Security measures	0.123	Reasonable secured (98% secured)	0
		Highly secured (99.99% secured)	100
Ease of use	0.147	Requires little knowledge	0
		Easy to use	100
Support	0.134	Phone and email	0
		Phone, live chat, video tutorials, and email	100
Cost	0.119	Expensive	0
		Affordable	100

enhance it further to provide a comparison interface for potential users of cloud services. This framework is a first effort in this context; in the future work, we are aiming to improve the model by exploring other relevant quality parameters by incorporate them into the model. Investigating the insights of cloud storage from different stakeholders such as ICT experts and cloud storage providers themselves is a future research opportunity.

7 Conclusion

Cloud storage is increasingly becoming important for many people. Currently, there are many cloud storage services offered by different Cloud providers. One of the challenges that are faced by cloud consumers is how to select the best cloud storage services which can satisfy their needs. Therefore, this paper proposed an approach for calculating the Cloud storage preference values based on criteria relative importance (weights). This paper presented a first multi parameters framework in evaluating cloud storage preferences.

References

1. Buyya, R., et al.: Cloud computing and emerging IT platforms: vision, hype, and reality for delivering computing as the 5th utility. Future Gener. Comput. Syst. **25**(6), 599–616 (2009)
2. Trustradius. Cloud Storage Providers (2015). https://www.trustradius.com/cloud-storage. Accessed 17 Mar 2016
3. GfK. How essential is cloud storage for global consumers? (2015). https://www.statista.com/statistics/552738/worldwide-cloud-storage-survey-importance-by-gender/. Accessed 10 Nov 2016
4. Cisco Visual Networking: Cisco Global Cloud Index: Forecast and Methodology, 2012–2017, (White Paper) (2013). http://www.cisco.com/c/dam/en/us/solutions/collateral/service-provider/global-cloud-index-gci/white-paper-c11-738085.pdf. Accessed 13 Nov 2016
5. Abran, A., et al.: Consolidating the ISO usability models. In: Proceedings of 11th International Software Quality Management Conference (2003)
6. Connor, D., et al.: Cloud storage: adoption, practice and deployment. An Outlook Report from Storage Strategies NOW (2011)
7. Walker, E., Brisken, W., Romney, J.: To lease or not to lease from storage clouds. Computer **43**(4), 44–50 (2010)
8. Ruiz-Alvarez, A., Humphrey, M.: A model and decision procedure for data storage in cloud computing. In: 2012 12th IEEE/ACM International Symposium on Cluster, Cloud and Grid Computing (CCGrid). IEEE (2012)
9. Garg, S.K., Versteeg, S., Buyya, R.: A framework for ranking of cloud computing services. Future Gener. Comput. Syst. **29**(4), 1012–1023 (2013)
10. ISO, ISO/IEC 9126: Information Technology-Software Product Evaluation-Quality Characteristics and Guidelines for Their Use (1991)
11. CSMIC, Service Measurement Index Framework Version 1.0 in USA 2014, Carnegie Mellon University Silicon Valley (2014)
12. Hansen, P., Ombler, F.: A new method for scoring additive multi-attribute value models using pairwise rankings of alternatives. J. Multi-criteria Decis. Anal. **15**(3–4), 87–107 (2008)
13. Forman, E.H., Selly, M.A.: Decision by Objectives: How to Convince Others that You are Right. World Scientific, Singapore (2001)
14. Saaty, T.L.: How to make a decision: the analytic hierarchy process. Eur. J. Oper. Res. **48**(1), 9–26 (1990)
15. Ombler, F., Hansen, P.: 1000Minds software (2012)
16. De Vaus, D.: Surveys in Social Research. Routledge, Abingdon (2013)
17. Macefield, R.: How to specify the participant group size for usability studies: a practitioner's guide. J. Usability Stud. **5**(1), 34–45 (2009)
18. Hastie, R., Dawes, R.M.: Rational Choice in an Uncertain World: The Psychology of Judgment and Decision Making. Sage, Thousand Oaks (2010)

Health Apps' Functionalities, Effectiveness, and Evaluation

Yazan Alnsour[1](✉), Bidyut Hazarika[2], and Jiban Khuntia[3]

[1] University of Illinois Springfield, Springfield, IL, USA
Yalns2@uis.edu
[2] Haworth College of Business, Western Michigan University,
Kalamazoo, MI, USA
bidyut.hazarika@wmich.edu
[3] Business School, University of Colorado Denver, Denver, CO, USA
Jiban.Khuntia@ucdenver.edu

Abstract. Positive user evaluation reflects success for a mobile application (app). When a health app fulfills their intended objective, it leads to higher usage and garners better evaluation. Although designing an app with a clear functionality is the key to success, but most apps are built with complex functionalities with confusing objectives that may not help a user's end objective of managing their health. In this regard, this study explores how functionality and intended health effectiveness of an app influence evaluation. Tracking a set of 188 health apps for 14 weeks, we find that functionality and appeal positively impact the evaluation of the apps. On the other hand, when the apps offer advanced and complex functionalities that are not mature and not fully integrated, the appeal will fade resulting in negative evaluations. Managerial and research contributions of the findings are discussed.

Keywords: Health apps · Apps functionalities · Mobile health · mHealth · Digital health

1 Introduction

Health information seeking has grown at a phenomenal rate. Health apps are programmed applications that run on smartphones and tablets to provide healthcare services. Health apps are intended to improve a patient's well-being. There are many types of health apps. Some apps are designed to help the patient manage his or her health, while others help patients and users to live a healthier lifestyle by giving them advice on nutrition and exercise, with others offer the ability for the patient to communicate with healthcare professionals regarding prescription refills and appointments [1, 2]. In addition, many apps are integrated with electronic medical records allowing health care providers to monitor patients and record their progress [3]. It is reported that there are 43,689 mobile healthcare apps available to users on the iTunes platform, with almost equal numbers in Android platform [4]. It is predicted that the number of users downloading health-related apps will reach to 1.7 billion by 2017 [5], with a global revenue potential of $21.5 billion by 2018 from mobile app based health businesses [6].

© Springer International Publishing AG 2017
M. Fan et al. (Eds.): WeB 2016, LNBIP 296, pp. 13–21, 2017.
https://doi.org/10.1007/978-3-319-69644-7_2

Research on the challenges of the health app usage and adoption suggest that apps should be designed properly for better appeal and usage, with appropriate design of their functionalities. The overall goal of a mobile health app is to improve the health and lifestyle of its users [7]. Mobile technology can be a powerful vehicle for providing individual level care to patients [8] if they provide effective functionalities. In this context, apps vary widely in the functionalities. Some apps offer instructive functionalities such as weight management, monitoring daily activities, patient reminders for pills and appointments, enabling patients to self-manage medical conditions, and locating nearby pharmacies and hospitals. Newer health apps offer more advanced functionalities such as monitoring vitals, communication with healthcare professionals and caregivers, ability to integrate with personal electronic health records and various other functions. However, with the objective to attract wider patient groups, often developers cram the apps with too many functionalities. While the intention of cramming too many functionalities may be to achieve a variety of tasks and objectives for the apps, but increasing functionalities may introduce complexity. The complexity and poor design may reduce overall appeal and lead to a negative impression and negatively impact subsequent usage intention [9]. Indeed, although health apps are perceived to be enablers toward personalized and patient centered care, yet, the realistic impact of apps on health is still low [10]. Given the potential benefits of health apps for both providers and patients, Kumar et al. [11] raise the need for more research in the area of health apps and the need for more collaboration between scholars from different disciplines like medicine, public health, and information systems areas.

We propose that the influence of some functionalities on the appeal and evaluation of a health app is amplified by the app effectiveness. In addition, the health effectiveness of an app moderates the effect of functionalities on appeal in such way that the appeal of the app decreases when the app offers advanced functionalities that are not fully developed. Furthermore, we propose that the influence of functionalities of an app on evaluation is also mediated by the appeal. In such, more appealing apps are rated higher by users. We tracked 188 mobile health apps in the Google digital market for 14 weeks. Text mining was used to code variables, and econometric techniques for panel data was used to analyze the data. This study contributes to the information systems and mobile health literature in providing a better understating of how functionalities impact the appeal and evaluation of the app.

2 Background and Prior Research

Phillips et al. [7] define mobile health as the usage of mobile technology to deliver and enhance healthcare experience to users. Related information systems and health information technology research have shown that health apps have the ability to deliver healthcare in an efficient way [11]. Health apps empower patients to manage their health, adopt a better lifestyle, and adhere better to recommended regimen [12]. Also, apps can enable patient self-monitoring [13] and provide healthcare professionals with additional intervention [14, 15] and treatment options [11, 12, 16–18].

Users can download health apps from the digital marketplace like itunes.apple.com for iOS devices, play.google.com for Android devices, and Microsoft.com store for

Microsoft devices [19]. The digital app markets provide users with a platform that allows them to review apps and rate them after they download and use the apps. The posted review and rating will influence other users' perception towards an app and may even impact the decision to download the app or not [20]. Recently, apps digital marketplaces allowed app developers to reply to reviews and comments. Such dynamics result in significant amount of unstructured data that provide researchers with a rich substance to explore and investigate.

To summarize, there has been an overload in the number of available health apps. Although there are many apps, developers and providers continue to encounter low usage and retention [21]. To increase the usage and the value of the apps, developers and providers tend to equip apps with more functionalities. Therefore, they need to better understand how adding more functionalities may impact users. To our knowledge, information systems and health IT researchers have not yet investigated how health apps functionalities may change user's evaluation. This study tries to address the existing gap in the literature of mobile health apps.

3   Theoretical Background and Conceptual Framework

In this research, we anchor to the prior research of assimilation and contrast effects on judgment and evaluation. The assimilation and contrast theory builds a foundation of an individual's evaluative judgment and attitude change. The theory argues that the addition of new information impacts an individual's initial opinion or attitude, the person will compare the new information to existing information and prior experience [22–24]. The theory explains how evaluation is based on a mental process that compares a given target to existing information, ideas, or experiences; this judgment is an accumulation of the assimilation and the contrast effects. The two processes are not mutually exclusive; while they may happen in separate of each other, they can also overlap depending on context and the situation. Based on this theory an individual may use a prior knowledge to evaluate a new experience. If the new target is not discrepant from the individuals existing expectations, the experience will be assimilated and falls in the area of acceptance. On the other hand, if the opposite is true then the new target will produce a contrast effect and cause user's rejection [23, 25, 26].

The assimilation in apps occurs when the target shares common features and attributes with existing or the previous ones. As opposed to that, the contrast happens when the experience falls outside the user knowledge or experience. The incorporation of the new experience that falls outside the user's expectation can be uncomfortable to the individual [23, 27]. Generally speaking, the assimilation and contrast effects are silent processes, and it is hard to identify their direction, but they are often reflected in the reviews, ratings, and attitudes towards products or services [28]. Assimilation effect is positive in nature where the contrast effect is negative [27]. Broadly, it is apparent that app developers need to stimulate more assimilation effect than contrast effect in their apps. Such will help in increasing the user's positive attitude and gain good evaluation from users.

We suggest three sets of relationships that will have an effect on the individual's overall evaluation of a health app. First, we propose that advanced and instructive functionalities will have a direct influence on the appeal and the evaluation of the app. The appeal in our study is the ability of the app to attract, please, interest, engage and enjoy users. Evaluation is reflected by the average rating an app gets every week. Second, we propose that the health app effectiveness, reflected by the positive impact noticed by the user due to using app, will moderate the impact of functionalities on the appeal and the evaluation of the app. Third, the appeal is a mediating factor between the role of functionalities and effectiveness on evaluation.

Based on the assimilation and contrast effects and our three suggested relationships, we argue that the foundation of a user's evaluation will be primarily drawn from the health app functionalities. For the health apps in this study, we operationalized a classification of health apps into advanced and instructive apps. The instructive functionalities are reflected through the features of providing information or instruction on the prevention of diseases, healthy living, self-diagnosis, finding a physician or facility, and post diagnosis education. The advanced functionalities are reflected through the features of providing reminders, alerts, connecting with healthcare professionals, teleconferencing, filling prescriptions, and compliance and adherence [4]. A number of functionalities on health effectiveness will be assimilated that will, in turn, enhance the user's evaluation and judgment. With more effectiveness, multi-functional apps will be able to attract more patients to realize the benefits of the app, and subsequently get higher evaluations. On the other hand, developing apps with advanced functionalities that are not mature or fully developed will result in a contrast effect that negatively impacts user's evaluation. In addition, appeal manages to establish a relationship between the app and the user. When the app is appealing, the user will evaluate the app positively and will be more tolerant towards glitches and bugs. We conduct an exploratory analysis of these effects using secondary data from the Android app store.

4   Method

4.1   Data and Variables

To test the proposed relationships, we focused on health apps in the Android app store. We collected data for a span of 14 weeks from October 2014 to January 2015. The first week is the focal reference week for the health apps in this study. We found more than 2,203 health and medical applications in the focal week of 13th October 2014 to 20th October 2014. We could not consider 49 apps for our analysis because they did not have any reviews in the marketplace in the focal week. After we had examined the data more closely, we found 73 apps in languages different than English, had unreadable names, or duplicated in the market. Also, we excluded 1,344 that were not related to patient's information, treatment, diagnosis, and or disease management. We also excluded 471 apps that were not directed to patients, rather directed towards providers, healthcare professionals, and medical students. We were left with 188 apps. Because this study spans for 14 weeks, we tracked these 188 apps for the 14 weeks to have an unbalanced (minimal) panel data set of 2,243 observations.

In Table 1, we provide a description of the variables we use in this study. To measure appeal and health effectiveness we used two different methods. In the first method, we examined the consistency between the polarity of review sentiment and the number of stars that a review is received, while in the second method, we used machine-learning techniques to predict the consistency of the review. Table 2 provides the descriptive statistics and correlations amongst key variables used in this study.

Table 1. Description of variables

Variable	Description and operationalization
Evaluation	Average weekly rating of the app in the app store
Functionalities	This variable is the total count of the four instructive functionalities and two advanced functionalities. The four major instructive functionalities of the apps: display of information, providing instructions, search and explore functions, and providing education. The instructive functionalities are reflected through the features of providing information or instruction on the prevention of diseases, healthy living, self-diagnosis, searching a physician or facility, and education on different procedures or conditions. The two advanced functionalities of the apps: connecting to back-end applications with the features and aligning to workflow and operational requirements. The advanced functionalities are reflected through the features of providing reminders, alerts, connecting or following up with doctors or providers, or with video or teleconferencing provisions, filling prescriptions, or compliance and adherence
Appeal	The ability of the app to attract, please, interest, engage and enjoy users in a way that will stimulate a relationship between the user and the app. This variable is coded by mining the text reviews of each app in each week
Health effectiveness	The positive impact felt or noticed by the user for using a health application. This variable is coded by mining the text reviews of each app in each week

Table 2. Descriptive statistics and pairwise correlations amongst key variables

Variable	Apps	Mean	Std. dev.	Min	Max	1	2	3	4	5	6	7	9
Evaluation	188	3.91	0.61	0	5	1							
Appeal	188	0.33	0.17	0	1	0.26	1						
Functionality	188	1.30	0.61	0	5	0.09	0.08	1					
Effectiveness	188	0.34	0.24	0	1	0.49	0.47	0.05	1				
Advanced	188	0.26	0.44	0	1	−0.01	−0.02	0.20	0.15	1			
Age	188	1.20	0.56	0.1	3.5	0.11	0.10	0.16	0.09	0.06	1		
Price	188	0.96	0.81	0	10	0.19	0.07	0.06	0.10	0.10	0.20	1	
Download	188	850	85	1	3.561	0.36	0.21	0.20	0.16	0.11	0.12	0.05	1

4.2   Estimation Models

We have used panel data for our analysis. To determine if we should perform a fixed effects or random effects analysis we ran Hausman test. The result of the test was significant, hence we used a fixed effect model. In addition, the change in patient evaluations is continuous which makes a fixed effect model a good fit for our analysis.

$$Evaluation_i = \beta X_{it} + \alpha + u_{it} + \varepsilon_{it}$$

where, $Evaluation_i$ is the dependent variable, X_i is a set of explanatory variables, β is a vector of parameters, t is the time in weeks, u is the between-entity error and ε are within entity error associated with each observation.

5   Results

We found that with more effectiveness, multi-functional apps will be able to better engage patients to use and form good relationship with the app. Also, with more appeal, multi-functional apps will be able to attract more patients to realize the benefits of the app, and subsequently get higher evaluations. In addition, we find support for the

Table 3. Results of Estimation Models

Variables	Fixed effect models			
	(1)	(2)	(3)	(4)
	Direct effect model	Direct effect model	Interaction model	Interaction model
	Appeal	Evaluation	Appeal	Evaluation
Effectiveness × functionalities			0.103*** (0.0121)	0.227*** (0.053)
Effectiveness × advanced			−0.110*** (0.013)	0.102*** (0.024)
Appeal		3.134*** (0.039)		3.100*** (0.039)
Health effectiveness	0.029*** (0.006)	0.022* (0.012)	0.031*** (0.009)	0.030* (0.017)
Functionalities	0.012* (0.015)	0.087*** (0.025)	0.011* (0.003)	0.053** (0.009)
Advanced	−0.023* (0.009)	−0.010* (0.016)	−0.194* (0.020)	−0.159* (0.038)
Constant	0.522*** (0.004)	2.251*** (0.022)	0.476*** (0.006)	2.229*** (0.022)
R-squared	0.310	0.347	0.341	0.381
F stat	15.85***	16.36***	16.48***	16.78***

(1) Significance levels: ***p < 0.01, **p < 0.05, *p < 0.10
(2) Standard errors in parentheses
(3) All Models have 188 number of apps, with 2,243 observations
(3) Models control for number of apps by same developer, developers' average ratings and reviews in the market, total rating of the developer, price of the app, when the app was last updated, when the app was introduced in the app store, number of downloads of the app, when the publisher released their first app in the app market

negative effect of advanced functionalities on the app appeal. The interaction term (Effectiveness x Advanced) is negative and significant (refer to column 3 in Table 3, $\beta = -0.110$, $p < 0.01$). The positive impact of effective functionalities on apps evaluation is supported as the coefficient is positive and significant (refer to column 4 in Table 3, $\beta = 0.102$, $p < 0.01$). Finally, appeal manages to establish a relationship between the app and the user. The user will evaluate the app positively and provide developers with better feedback to improve it. Users will also be more tolerant towards glitches and bugs in apps that they find appealing.

We tested for multicollinearity by computing variance inflation factors (VIFs) for all estimation models. The highest VIF was 2 in the direct-effect models, confirming that multicollinearity is not a serious concern. To reduce potential high multicollinearity issues due to the number of interaction terms in the models, all continuous variables were mean-centered by subtracting the corresponding variable mean from each value [29]. The VIF of any individual variable in any of the interaction effect models was less than 7. Furthermore, mean VIFs in all the models were less than 5. Thus, we find that multicollinearity is not a serious concern in the estimation.

6 Discussion

The first thing we can draw from our findings is that adding more effective functionalities to health apps help increase the average rating. Secondly, adding advanced functionalities in the health app may sacrifice some of its simplicity an appeal that will negatively impact the overall rating of the app. A third finding is that the appeal of the health app will positively impact its average rating.

We draw some managerial implications from this study. First, the apps functionality plays a valuable role in user's evaluation. Hence, developers should pay attention toward what type of functionalities they provide in their applications. This study contributes to the literature of mobile health applications, by identifying how technological and functional factors are associated with digital application success. Finally, tracking evaluation regularly is critical for a health application's success.

Regarding our research contributions, to our knowledge, this study will be the first one to explore the effects of factors like functionality, appeal, and effectiveness on how users rate health apps in the digital marketplace. These contributions will enrich the existing information systems literature and research associated with mobile health apps. Future studies can look if health apps were recommended or prescribed by providers and how that will impact consumer's evaluation. In our study, we only collected data for apps that came from the Android store; future studies may include apps from other markets as well.

In conclusion, as mobile health apps become ubiquitous, Health apps functionalities and appeal will have a stronger relationship with users' initial and long-term decision to use the app. Our study shows that developers should pay attention not only to health app's functionalities but also how appealing the app is. Developers tend to add more functionalities to their apps to keep up with competition sometimes at the expense of simplicity and appeal. Developers that contradict what users find appealing may trigger a bad first impression and damage user's expectations. As functionality is important to establish a long-term relationship between the user and the health app, the appeal is significant in leaving a good impression to start using the app and engage users more.

References

1. Boulos, M.N.K., et al.: Mobile medical and health apps: state of the art, concerns, regulatory control and certification. Online J. Public Health Inform. **5**(3), 229 (2014)
2. Fox, S., Duggan, M.: Mobile Health 2012. Pew Internet & American Life Project, Washington (2012)
3. Lobelo, F., et al.: The wild wild west: a framework to integrate mHealth software applications and wearables to support physical activity assessment, counseling and interventions for cardiovascular disease risk reduction. Prog. Cardiovasc. Dis. **58**, 584–594 (2016)
4. Aitken, M., Gauntlett, C.: Patient apps for improved healthcare: from novelty to mainstream. Parsippany NJ: IMS Inst. Healthc. Inform. (2013)
5. Economist: Mobile health apps are becoming more capable and potentially rather useful (2016). http://www.economist.com/news/business/21694523-mobile-health-apps-are-becoming-more-capable-and-potentially-rather-useful-things-are-looking
6. BCC: Global mHealth Technologies Market Projected to Reach nearly $21.5 Billion in 2018; Europe Segment Growing At 61.6% CAGR (2014). http://www.bccresearch.com/pressroom/hlc/global-mHealth-technologies-market-projected-to-reach-nearly-$21.5-billion-2018
7. Phillips, G., et al.: The effectiveness of M-health technologies for improving health and health services: a systematic review protocol. BMC Res. Notes **3**(1), 250 (2010)
8. Free, C., et al.: The effectiveness of mobile-health technology-based health behaviour change or disease management interventions for health care consumers: a systematic review. PLoS Med. **10**(1), e1001362 (2013)
9. Tuch, A.N., et al.: The role of visual complexity and prototypicality regarding first impression of websites: working towards understanding aesthetic judgments. Int. J. Hum.-Comput. Stud. **70**(11), 794–811 (2012)
10. Wu, L., Li, J.-Y., Fu, C.-Y.: The adoption of mobile healthcare by hospital's professionals: an integrative perspective. Decis. Support Syst. **51**(3), 587–596 (2011)
11. Kumar, S., et al.: Mobile health technology evaluation: the mHealth evidence workshop. Am. J. Prev. Med. **45**(2), 228–236 (2013)
12. Free, C., et al.: The effectiveness of mobile-health technologies to improve health care service delivery processes: a systematic review and meta-analysis. PLoS Med. **10**(1), e1001363 (2013)
13. Ramanathan, N., et al.: Identifying preferences for mobile health applications for self-monitoring and self-management: focus group findings from HIV-positive persons and young mothers. Int. J. Med. Inform. **82**(4), e38–e46 (2013)
14. Klasnja, P., Pratt, W.: Healthcare in the pocket: mapping the space of mobile-phone health interventions. J. Biomed. Inform. **45**(1), 184–198 (2012)
15. Terry, M.: Medical apps for smartphones. Telemed. J. EHealth **16**(1), 17–22 (2010)
16. Dahne, J., Lejuez, C.W.: Smartphone and mobile application utilization prior to and following treatment among individuals enrolled in residential substance use treatment. J. Subst. Abuse Treat. **58**, 95–99 (2015)

17. Gustafson, D.H., et al.: A smartphone application to support recovery from alcoholism: a randomized clinical trial. JAMA Psychiatry **71**(5), 566–572 (2014)
18. Milward, J., et al.: Mobile phone ownership, usage and readiness to use by patients in drug treatment. Drug Alcohol Depend. **146**, 111–115 (2015)
19. Sunyaev, A., et al.: Availability and quality of mobile health app privacy policies. J. Am. Med. Inform. Assoc. **22**(e1), e28–e33 (2015)
20. Huang, G.-H., Korfiatis, N.: Trying before buying: the moderating role of online reviews in trial attitude formation toward mobile applications. Int. J. Electron. Commer. **19**(4), 77–111 (2015)
21. Rai, A., et al.: Understanding determinants of consumer mobile health usage intentions, assimilation, and channel preferences (2013)
22. Johnson, M.D., Fornell, C.: A framework for comparing customer satisfaction across individuals and product categories. J. Econ. Psychol. **12**(2), 267–286 (1991)
23. Förster, J., Liberman, N., Kuschel, S.: The effect of global versus local processing styles on assimilation versus contrast in social judgment. J. Pers. Soc. Psychol. **94**(4), 579 (2008)
24. Walker, J.L.: Assimilation/contrast effects: the effect of expectations across the goods/services continuum (1994)
25. Hovland, C.I., Harvey, O., Sherif, M.: Assimilation and contrast effects in reactions to communication and attitude change. J. Abnorm. Soc. Psychol. **55**(2), 244 (1957)
26. Stapel, D.A., Koomen, W., Velthuijsen, A.S.: Assimilation or contrast?: Comparison relevance, distinctness, and the impact of accessible information on consumer judgments. J. Consum. Psychol. **7**(1), 1–24 (1998)
27. Stewart, K.J., Malaga, R.A.: Contrast and assimilation effects on consumers' trust in Internet companies. Int. J. Electron. Commer. **13**(3), 71–94 (2009)
28. Oliver, R.L., Burke, R.R.: Expectation processes in satisfaction formation a field study. J. Serv. Res. **1**(3), 196–214 (1999)
29. Aiken, L.S., West, S.G., Reno, R.R.: Multiple Regression: Testing and Interpreting Interactions. Sage, Newcastle upon Tyne (1991)

Predicting Online Reviewer Popularity:
A Comparative Analysis of Machine Learning
Techniques

Samadrita Bhattacharyya[✉], Shankhadeep Banerjee,
and Indranil Bose

Management Information Systems, Indian Institute of Management Calcutta,
Calcutta, India
{samadritab14,shankhadeepb15,bose}@iimcal.ac.in

Abstract. Online customer reviews have been found to vary in their level of
influence on customers' purchase decisions depending on both review and
reviewer characteristics. It is logical to expect reviews written by popular
reviewers to wield more influence over customers, and therefore an investigation
into factors which can help explain and predict reviewer popularity should have
high academic and practical implications. We made a novel attempt at using
machine learning techniques to classify reviewers into high/low popularity
based on their profile characteristics. We compared five different models, and
found the neural network model to be the best in terms of overall accuracy
(84.2%). Total helpfulness votes received by a reviewer was the top determinant
of popularity. Based on this work, businesses can identify potentially influential
reviewers to request them for reviews. This research-in-progress can be extended using more factors and models to further enhance the accuracy rate.

Keywords: Online reviews · Reviewer popularity · Machine learning
techniques · Predictive analytics

1 Introduction

Widespread access to the internet is changing the way modern consumers make their
purchase decisions. This change is facilitated by e-commerce platforms like Amazon
and other online review websites like Yelp which host customer reviews of various
products and services. Online customer reviews, also known as electronic
word-of-mouth (eWOM) can be defined as "*any positive or negative statement made by
potential, actual, or former customers about a product or company, which is made
available to a multitude of people and institutions via the Internet*" [1]. According to a
survey[1], for 90% of customers, their buying decisions were influenced by online
reviews. Academic studies too have confirmed the importance of online reviews on
customers' purchase decisions [2]. However, consumers find some reviews more

---

[1] http://marketingland.com/survey-customers-more-frustrated-by-how-long-it-takes-to-resolve-a-customer-service-issue-than-the-resolution-38756

.

© Springer International Publishing AG 2017
M. Fan et al. (Eds.): WeB 2016, LNBIP 296, pp. 22–28, 2017.
https://doi.org/10.1007/978-3-319-69644-7_3

helpful than the others. This could be either because of the characteristics of the review content (review length, review polarity, content and style) [3, 4] and/or the characteristics of the reviewer.

It has been found that apart from information quality, source credibility is an important aspect for information seeking and adoption [5, 6]. In the context of online reviews, 'source' would imply the reviewer, hence reviewer credibility should impact review adoption. An accumulation of reviewer credibility over time should lead to more customers following a reviewer, leading to more reviewer popularity. Thus, popularity of reviewers should be associated with the impact of their reviews on the customers. The importance of word-of-mouth of influential reviewers on consumers' purchase decision is well established in previous research [7, 8]. However, little attention has been given to study factors which make reviewers popular, and hence more influential. These factors could be used to predict reviewer popularity, which could be useful for businesses. Hence, in this research-in-progress, we attempt to identify popular reviewers based on their online profile characteristics. We use data from Yelp website which has an extensive reviewer community and detailed reviewer attributes. One of such attributes is the number of followers of a reviewer, which we use as a proxy for popularity. Other information regarding the reviewer include the number of reviews written, the number of friends a reviewer has, average rating the reviewer provides, years of experience in writing reviews for the website, etc.

We used five different machine learning techniques to classify reviewers as high or low on popularity based on their profile characteristics provided, and compared their performances. Also, we identified the factors which were most impactful in predicting reviewer popularity.

Insights shared in this study might help businesses in targeting popular reviewers for writing reviews about their offerings. Again, by predicting popularity of a reviewer, review websites might prioritize the display of a new review which is yet to get a helpfulness vote. The study also contributes to the growing research on online customer reviews and to the best of our knowledge, this is a novel attempt of using predictive analytics in the context of reviewer popularity.

2 Literature Review

Past literature one-WoM has primarily focused on identifying the factors related to helpfulness of reviews [9, 10]. A study by [11] found that perceived value of review is influenced by reviewer's expertise and reputation. [10] used average helpfulness votes received per review and personal information disclosure for finding impact on review helpfulness. Another research has found that reviewer characteristics like the number of reviews posted by a reviewer and the number of helpful votes received by the reviewer on the whole, impacts the helpfulness vote of a review [12]. Study also found that reviews written by a self-described expert are more helpful than those that are not [13]. Some reviewer characteristics such as reviewer quality and reviewer exposure are found to impact sales by reducing perceived uncertainty of buyers [14]. However, to the best of our knowledge there has been no research to identify the dominant factors responsible for making a reviewer popular.

In our study we attempt to differentiate relatively more popular and less popular reviewers using a predictive analytics approach. The determining factors are selected based on support from extant literature and availability of data. Since we did not find much literature directly examining the factors related to reviewer popularity, so we had to use nearest available proxy which is review popularity to justify our variables. Review popularity is based on the perceived helpfulness of the reviews. Review helpfulness has been found to be influenced by reviewer characteristics [13, 15, 16], and thus justifies its use as the proxy. Hence, for predicting reviewer popularity, we identify factors from literature which influence helpfulness of the reviews.

[17] using TripAdvisor data studied the effect of review polarity on helpfulness of the review. They found that reviewers who posted more positive reviews are more likely to receive helpful votes than those who stressed on the negative aspects. Star rating is an indicator of reviewer's polarity. Hence, we include average review rating as a factor of reviewer popularity.

Another important factor found to be influential in deciding review helpfulness is review length [4, 18, 19]. Studies have found that review length provides important cues regarding reviewer characteristics [20]. Prior literature has already established that number of reviews are associated with review helpfulness [11, 12]. Drawing on these we can say that more the number of reviews a user writes on a forum, more popularity she gains.

Also, it could be said that with increased experience, a reviewer writes more useful reviews and hence become more popular. Similarly, the helpfulness votes a reviewer receives should be associated with her popularity, since it validates her credibility. The research by [10] also confirms that average helpfulness votes received by reviewer as one of the possible reviewer characteristics that might affect review helpfulness.

Being a well-reputed reviewer with certification from the website ('Elite' in case if Yelp website) also establishes the reviewer credibility and in turn should influence popularity. Finally, more the number of friends a reviewer has, more popular she is expected to be [11] used number of friends as a measuring variables for reviewers' reputation, hence we use number of friends as a factor of reviewer popularity.

3   Data and Methodology

A large dataset with 552,339 records was collected from Yelp.com which was made public as a part of the Yelp Dataset Challenge 2016. After processing data and removing outliers we had 69,612 records which we used for analysis. Yelp hosts customer reviews on local businesses. The reason behind selecting the website is that it provides information regarding the reviewers and their followers. User attributes such as number of followers, number of friends, average review rating, number of reviews written, total helpfulness votes, years of experience, years of reputation, and average review length for each reviewer were provided in the data. Number of followers was used as a proxy for reviewer popularity. Description on the data is shown in Table 1.

Table 1. Data descriptive

Variable	Range	Mean	SD
Number of followers	1–23	2.042	1.934
Average review rating per user	1–5	3.816	0.587
Number of reviews written	3–284	45.026	46.541
Years of experience	1–12	5.214	2.084
Years of reputation	0–3	0.133	0.435
Average review votes per user	0–8.5	2.085	1.414
Number of friends	1–49	7.186	9.361
Average review length	1–58	23.114	13.130

We used clustering technique (2-stage clustering) to decide on the number of segments appropriate for classification. The results showed two distinguished clusters. On observation of the clustered data we found mean value of number of followers of a reviewer to be the demarcation value. The reviewers having followers more than mean value are said to be high on popularity and vice-versa. The outcome variable is binary with 1 representing high and 0 representing low.

Data was partitioned in 70:30 ratio for training and testing. Five different models were used for classification: C5, Neural network, Bayesian network, CHAID, and Logistic Regression. We used IBM SPSS modeler as the analytical tool. The models were compared based on overall accuracy, lift, and costs. The agreement of all the models were checked to ensure their comparability.

4   Results and Analysis

Different models had different values of accuracy, however, they didn't differ much. The overall accuracy was around 83%–84%. There was 83.8% agreement among the classification techniques. Table 2 summarizes the results.

Table 2. Summary of results for various predictive models

Models	Overall accuracy	Lift	Factors used	AUC
Neural network	84.2%	2.43	7	0.863
Logistic regression	83.82%	2.39	7	0.856
C5 1	83.62%	2.33	7	0.810
Bayesian network	83.52%	2.33	7	0.822
CHAID 1	83.38%	2.39	5	0.856

All the models show nearly same level of accuracy with neural network giving the best value. We found that number of reviews and average helpfulness votes received by a reviewer were the two most important predictors among all followed by number of friends and average review rating. The least important factors turned out to be average review length and years of experience. Figure 1 depicts the predictor importance. Table 3 shows the confusion matrix for neural network. 85.9% of reviewers who are low on popularity are predicted correctly, whereas 70.1% of reviewers high on popularity were correctly predicted. Prediction accuracy is higher for less popular class. Businesses would try to minimize the number of less popular reviewers being predicted as more popular ones, since that would incur costs in investing their time and resources on uninfluential reviewers. In our model, this case was found to be just 14.1%, which is on the lower side of error.

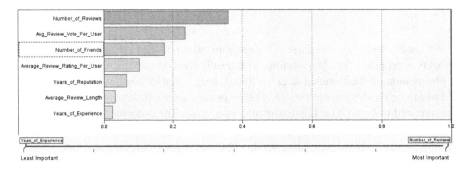

Fig. 1. Predictor importance graph

Table 3. Confusion matrix for neural network

Observed	Predicted	
	0	1
0	85.9%	14.1%
1	29.9%	70.1%

All other models except CHAID used all of the inputs to predict the output variable. CHAID model discarded years of 'Elite' and average review length as predictors.

5   Conclusion

In this research-in-progress paper, we attempted to use predictive analytics to classify online reviewers into two distinct classes based on their popularity. We compared five different machine learning techniques - C5, Neural network, Bayesian network, CHAID, and Logistic Regression. Among all, the neural network model turned out to be the best with 84.2% accuracy. Number of reviews written was found to be the most important factor.

In future we plan to incorporate few more factors such as review content characteristics, and try to improve the accuracy of prediction. For example, review subjectivity, polarity, topic relevance, spelling & grammar, etc. could be some more variables to consider. Additionally, we also want to create ensemble models using more than one predictive models (like using a neural network for accuracy and a decision tree for rules), and analyze the results.

Owing to the significant impact of online reviews on customers' purchase decisions, it is important for the businesses to manage the reviews received for their products and services. In order to get more impactful reviews, it is important for businesses to identify influential and popular reviewers. Based on some characteristics or cues about the reviewer, if it is possible to predict reviewer popularity, businesses could leverage the information to target those reviewers and encourage them to write reviews about their products or services. If necessary, they might also incentivize the most popular reviewers. Also, it is advisable for the businesses to keep a track of the issues raised by popular reviewers to proactively address those. Businesses could use extract ideas from their reviews to enhance the offerings if needed.

For e-commerce and online review sites, the insights from this study could be helpful in many ways. They can develop recommender systems based on different characteristics of a reviewer, predict their popularity, and display their reviews as top reviews on their sites. This would be particularly useful for those websites where social interaction (like following) among reviewers and other consumers is not possible.

References

1. Hennig-Thurau, T., Gwinner, K.P., Walsh, G., Gremler, D.D.: Electronic word-of-mouth via consumer-opinion platforms: what motivates consumers to articulate themselves on the internet? J. Interact. Mark. **18**(1), 38–52 (2004)
2. Zhu, F., Zhang, X.: Impact of online consumer reviews on sales: the moderating role of product and consumer characteristics. J. Mark. **7**(2), 133–148 (2010)
3. Korfiatis, N., García-Bariocanal, E., Sánchez-Alonso, S.: Evaluating content quality and helpfulness of online product reviews: the interplay of review helpfulness vs. review content. Electron. Commer. Res. Appl. **11**(3), 205–217 (2012)
4. Mudambi, S.M., Schuff, D.: What makes a helpful online review? A study of customer reviews on Amazon.com. MIS Q. **34**(1), 185–200 (2010)
5. Cheung, C.M.K., Lee, M.K.O.: Information adoption in an online discussion forum. In: 2nd International Conference on e-Business, ICE-B 2007, pp. 322–328 (2007)
6. Zhang, W., Watts, S.: Knowledge adoption in online communities of practice. In: ICIS 2003 Proceedings, vol. 9(1), pp. 96–109 (2003)
7. Kim, Y., Srivastava, J.: Impact of social influence in e-commerce decision making.In: Proceedings of ninth International Conference on Electronic Commerce - ICEC *2007*, p. 293 (2007)
8. Li, Y.M., Lin, C.H., Lai, C.Y.: Identifying influential reviewers for word-of-mouth marketing. Electron. Commer. Res. Appl. **9**(4), 294–304 (2010)
9. Devi, J.: Estimating the helpfulness and economic impact of product reviews. Int. J. Innov. Res. Dev. **1**(5), 232–236 (2012)
10. Ghose, A., Ipeirotis, P.G.: Estimating the helpfulness and economic impact of product reviews: mining text and reviewer characteristics. IEEE Trans. Knowl. Data Eng. **23**(10), 1498–1512 (2011)

11. Liu, Z., Park, S.: What makes a useful online review? Implication for travel product websites. Tour. Manag. **47**, 140–151 (2015)
12. Otterbacher, J.: Helpfulness' in online communities: a measure of message quality. In: Proceedings of the SIGCHI Conference on Human Factors in Computing Systems, pp. 1–10 (2009)
13. Connors, L., Mudambi, S.M., Schuff, D.: Is it the review or the reviewer? A multi-method approach to determine the antecedents of online review helpfulness. In: Proceedings of the 44th Hawaii International Conference on System Sciences, pp. 1–10 (2011)
14. Hu, N., Liu, L., Zhang, J.: Do online reviews affect product sales? The role of reviewer characteristics and temporal effects. Inf. Technol. Manag. **9**(3), 201–214 (2008)
15. Forman, C., Ghose, A., Wiesenfeld, B.: Examining the relationship between reviews and sales: the role of reviewer identity disclosure in electronic markets. Inf. Syst. Res. **19**(3), 291–313 (2008)
16. Ngo-Ye, T.L., Sinha, A.P.: The influence of reviewer engagement characteristics on online review helpfulness: a text regression model. Decis. Support Syst. **61**(1), 47–58 (2014)
17. Fang, B., Ye, Q., Kucukusta, D., Law, R.: Analysis of the perceived value of online tourism reviews: influence of readability and reviewer characteristics. Tour. Manag. **52**, 498–506 (2016)
18. Kim, S.-M., Pantel, P., Chklovski, T., Pennacchiotti, M.: Automatically assessing review helpfulness. In: Proceedings of 2006 Conference Empirical Methods Nature Language Processing (EMNLP 2006), pp. 423–430, July 2006
19. Schindler, R.M., Bickart, B.: Perceived helpfulness of online consumer reviews: the role of message content and style. J. Consum. Behav. **11**(3), 234–243 (2012)
20. Baek, H., Ahn, J., Choi, Y.: Helpfulness of online consumer reviews: readers' objectives and review cues. Int. J. Electron. Commer. **17**(2), 99–126 (2012)

Amazon and Alibaba: Competition in a Dynamic Environment

Wei-Lun Chang[1][(✉)] and Thomas J. Allen[2]

[1] Tamkang University, New Taipei City, Taiwan
wlchang@mail.tku.edu.tw
[2] Naitonal Sun-Yat Sen University, Kaohsiung City, Taiwan
thomasjhallen@gmail.com

Abstract. This paper investigates the dynamic nature of the e-commerce industry and the competitive relationships based on the market commonality and resource similarity of a focal firm. This theoretical investigation is prefaced by a critical analysis of the factors that contributed to the birth of the digital revolution and its subsequent effect on commerce. The methodological process used is called competitor mapping, which is an extension of the awareness-motivation-capability framework. This study discovered that companies that are characterized by having low market commonality and low resource similarity tend to engage in competition at a moderate level. Strategic decisions or actions taken are typically in the form of attempts to enter new markets or increase current market shares. In this study, each firm's respective counterpart dominates a valued market, which elicits an opportunity for simultaneous cooperation and competition.

Keywords: Competitive dynamics · Cwareness-motivation-capability · Competitor mapping · Coopetition · Competition

1 Introduction

There has been an unprecedented shift in the way people use the Internet in the last two decades. Globalization has accelerated due to the growing ubiquity of the Internet. As a result, society has developed a progressive way of thinking about how people and businesses interact with one another along the supply chain. Hunter (2013) states that many inventions, their ensuing commercialization, and their acceptance by society have dramatically changed our way of life. Society owes the progression of our social existence to the invention of new technologies. In the initial stages of this technological development, computers were once the size of entire rooms; now pieces of technology of exponentially greater computing power fit inside one's pants pocket (Alba 2015). Moore's law states that the growth and evolutionary possibilities of technological computing power have the potential to be exponential. The vastness of e-commerce in the modern era is a direct result of these revolutionary ideas of the past.

The impact that the Internet and e-commerce have on our lives will only continue to grow. The e-commerce phenomenon is an organic ecosystem consisting of enterprises and organizations with close relations that use the Internet as a platform to engage in

© Springer International Publishing AG 2017
M. Fan et al. (Eds.): WeB 2016, LNBIP 296, pp. 29–53, 2017.
https://doi.org/10.1007/978-3-319-69644-7_4

competition and communication through virtual alliance, thereby sharing resources and making full use of their advantages beyond geographical limits (Huang et al. 2009). The key here is to identify the competitive nature of businesses in the e-commerce space. Companies such as eBay, Amazon, Alibaba Group, JD.com, and Rakuten have emerged as dominant players in this space, solidifying their hold in their respective markets around the globe. The online retail space is dynamic and constantly in flux, which makes strategic decision-making and inter-firm rivalry a complicated issue. When the majority of methods of traditional differentiation become homogenous across all firms, the only method of differentiation is the price point. While deploying the Internet can expand the market, doing so often comes at the expense of average profitability.

By virtue of providing products and services online, e-commerce companies are not restricted to single locations, unlike traditional businesses. From the perspective of the consumer, this reduces the cost of participating in a market as well as the physical and mental effort it requires (Laudon 2014). When given a choice, people will always choose the path that requires the least amount of effort: the most convenient path (Shapiro and Varian 1999). In an attempt to understand these competitive relationships between major players in e-commerce better, this study examines the unique features of the e-commerce industry, and some behavioral aspects of these companies during competition and the strategic and operational decisions made by them.

E-commerce is commonly associated with high levels of growth and expansion into new markets. Each player within the industry is able to remain relevant and exploit its competitive advantage. It is difficult to gain competitive advantage in such a dynamic space. One way to attain this advantage is for companies to grow as quickly as possible and establish a presence, which leads to user loyalty. This presence is critical to limit the ability of competitors to enter the market and to exploit existing ideas in better ways (Hunter 2013). The trend has continued to be true for the modern e-commerce business strategy.

The industry leaders set the standard for all others who operate in their market. Best practice competition eventually leads to competitive convergence, with many companies doing the same things in the same way (Porter 2001). Presently, the society is not adequately informed about the aforementioned phenomena of e-commerce. The flow of information is quite one sided, and an in-depth analysis will be conducted in this study in an attempt to understand this industry better. This paper explores how two fundamental factors of competition affect the strategy of firms in online marketplaces and their implications in the context of the two most important firms in the e-commerce space: Amazon.com, Inc. (hereafter Amazon), and Alibaba Group Holdings Limited (hereafter Alibaba). This study explores three distinct research questions:

RQ1: In what way has the development of technology influenced the nature of business?
RQ2: How do the strategic positions of firms shape the competitive relationships?
RQ3: In what way(s) do a firm's competitive relationships affect other firms?

These broad questions investigate a spectrum of information that provides a better understanding of the impact of information technology on the business environment, the characteristics of this newly formulated business environment, and the way in which major players interact with one another within this unique business environment. This paper investigates the nature of the industry by focusing on dominant firms. The selection of these dominant firms is directly related to their influence and power within the industry. Technology has fundamentally changed the way society interacts and this is highlighted perfectly by the creation of an entire business world online. With that in mind, this paper describes and characterizes the online industry through the application of relevant theoretical frameworks.

The theory of competitive dynamics has been used in existing research, mainly in the airline industry, to study the establishment of alliances among competitors. This theory examines how different organizations vie for dominance in a certain space. To be more specific, Chen and Miller (2012) define competitive dynamics as the study of inter-firm rivalry based on specific competitive actions and reactions, strategic and organizational contexts, and drivers and consequences. The theory serves as a synthesizing framework for linking strategic content and process, resource-based and market perspectives, and strategy development and implementation.

An integral aspect of this competitive theory is the Awareness-Motivation-Capability (AMC) model. This model allows us to conduct a critical analysis of both macro-competitive and micro-actor viewpoints. At its fundamental core, the AMC model characterizes three main aspects of a firm's competitive activity: first, its *awareness* of other firms' moves within a shared space; second, its *motivation* to respond to a given action; and third, its *capability* to respond. In addition to the theoretical application of the competitive dynamics theory, this paper also provides a preliminary characterization of the modern e-commerce industry in the United States and the Asia-Pacific region through the lens of Michael E. Porter's Five Forces model of competition.

Furthermore, this paper presents the competitor mapping of Amazon and Alibaba. Competitor mapping is a methodological process that creates a visualization of the competitive tensions associated with rival firms in a shared industry. Competitive tension may be a strategic approach characterized by its asymmetrical nature. This competitive asymmetry phenomenon often appears across various industries (DeSarbo et al. 2006). Despite this asymmetry, a dyadic relationship is revealed through the collection of specific financial and operational datasets and the use of formulas for each company. This relationship clarifies the affiliation and level of competition between these two firms in the global market place.

2 Literature Review

Competitive analyses related to the AMC model have been employed to gain insight into the intricacies of various industries. Peng et al. (2012) used this theory to investigate the effects of competition and coopetition in the Taiwanese supermarket industry. Owing to the complex nature of coopetition, the study was conducted over a fifteen-year period. The study analyzed the performance of a firm before and after coopetition became the norm in the industry. The findings revealed that this kind of coopetition leads to better performance over time in two ways. First, the adaptation of coopetition allows companies to reach a level of performance that would otherwise be unreachable. Second, the implementation of these cooperative principles allows this performance level to be attained at an earlier point in time; therefore, we conclude that, albeit temporarily, coopetition within a competitive environment yields advantageous results.

Chen et al. (2007) use the AMC model to investigate the competitive tension between rival firms in the airline industry. The application of AMC theory to the airline industry is common. The highly competitive nature of the industry and the establishment of alliances make it an ideal environment for research. By developing a statistical approach, they draw several conclusions based on the AMC model and other theories of competition. They find that each firm's perceived tensions vary from one competitor to another. A competitor's degree of rivalry is not equal and in fact changes from firm to firm. This variability in perceived tension could act as a catalyst for action and response or encroachment on a rival's market share. Although perceived tension can be characterized as asymmetrical across the participating firms, the perception of these tensions is not comprehensive. Managers control the aspirations of a firm, but stakeholders may have a greater influence on the industry as a whole.

Gnyawali et al. (2006) argue that a firm's structural position within an industry affects its degree of strategic coopetition and competitive resource allocation. Resource allocation depends on the degree to which resource bundles held within a firm are alike or asymmetrical. The classification of a firm's resource bundles has a direct impact on its outward competitive strategy. Resource bundles can be characterized as human, financial, or knowledge capital. Gnyawali et al. (2006) focus on the global steel industry because of the diversity and structural variance across the firms in this industry. They deduce that the less firms have in common in term of resources, the higher is the frequency of action. Moreover, highly centralized firms are involved in a greater number of competitive actions with other rivals in the industry.

Haleblian et al. (2012) investigate how the awareness, motivations, and capabilities of firms affect their success during merger activity. In addition, their paper also takes into account the sequence of actions taken by these firms while relating it to game theory and the first mover advantage. They characterize a firm's strategic emphasis, structure, and resource allocation along with its effect on participation and success in periods of high merger activity. They conclude that firms that experience high levels of growth, lower debt/equity ratios, and higher levels of efficiency tend to participate more frequently in acquisition activities. The early movers are more aware, have greater motivation to act, and have superior capabilities. This paper therefore suggests that firms should improve their detailed knowledge of other firms, the industry itself, and the environment in which they operate. Firms should also stringently assess perceived gains and losses from possible actions and focus on the thoughtful deployment of resources to become an early mover in periods of high merger activity and increase their overall success rate.

Tsai (2002) discusses the competitive environment within organizations. This study discusses the same competitive environment within a single organization. The propensity of employees of the same company to engage in fierce competition is not well-known. The study also identifies different organizational structures and their propensity to elicit competitive or cooperative tendencies in terms of knowledge sharing and mutually beneficial relationships. Highly centralized firms structured as formal hierarchies typically promote a more competitive environment among these intra-organizational units. This typically has a negative effect on lateral relations. Informal or decentralized organizations typically promote cooperation, which in turn, can elicit coopetition, which Peng et al. (2012) conclude is a positive outcome for business. The actions of these internal stakeholders (managers) provoke positive responses from the employees of the firm.

Baum and Korn (1999) examine the relationships involved in the entries and exits of firms into each other's markets. The study uses a dyadic approach to investigate the actions and responses for the companies making decisions to enter these new markets. It links the concepts of multimarket competition and cooperation with inter-firm rivalry and the asymmetrical perceptions each firm has with one another. This study also analyzes the airline industry due to the fiercely competitive attitudes of airline companies and the tendencies of firms within that industry to move in and out of different markets frequently. They conclude that a firm's overall performance varies greatly depending on the continued interaction of a firm and its direct rivals. This interaction is evident in the actions and responses of each firm and how these actions or responses are directed at competitors.

Chen (2008) considered the paradoxical relationship between competition and cooperation typically understood in the field of strategy and business. There is a significant amount of research linked with the idea that these two ideas are systematically contradictory and therefore involve little to no interplay. Chen (2008) uses the opposing ideological norms of western and eastern culture to highlight that not only do these concepts have interrelated aspects, but the paradoxical nature of these two themes in fact facilitate further investigation and exploration. Chen (2008) notes that, while on the surface, cooperation and competition may seem to represent two contradictory ideas, when blended together, they create the idea of coopetition, which is an increasingly popular notion in business strategy. Therefore, the continued research of this correlation is imperative.

Some concepts in the literature are the fundamental foundations of competitive theory. These concepts have been employed by a variety of scholars, all of whom explore the details of strategic interaction among firms in a shared industry. While not all concepts are used in every study, most studies apply some combination of concepts best suited to the study itself. To understand the concepts in the existing literature better, Table 1 illustrates the theoretical focus of each paper (Table 1).

Table 1. Common concepts in the literature

	Concepts			
	Coopetition	Resource similarity	Market commonality	Action/response dyad
(Peng et al. 2012)	V	V	V	
(Chen et al. 2007)				V
(Gnyawali et al. 2006)	V	V		V
(Haleblian et al. 2012)			V	V
(Tsai 2002)	V			V
(Baum and Korn 1999)	V		V	V
(Chen 2008)	V			

These four concepts have been highlighted because of their immediate relevance to this specific study. Within the e-commerce space, this paper investigates the impact of the two fundamental concepts of AMC competitive theory—resource similarity and market commonality. The AMC model allows us to incorporate all these concepts to study the competitive nature of the industry. Chen (1996) and Chen and Miller (2012) state that awareness is considered a prerequisite for any move that would include the decision to cooperate with a competitor. The likelihood of cooperation will increase with a firm's market commonality and resource similarity with its competitors. The motivation of a firm to attack or respond to an attack is affected by the degree of its market commonality with that of its competitor, and resource similarity is important for determining the capability of a firm to attack or respond. Evidently, these fundamental concepts are interrelated and provide insight into the nature of competition. Coopetition is alluded to as well because its feasibility depends on how two firms perceive one another based on the market and resource constructs.

3 Methodology

3.1 Content Analysis

This paper analyzes the effect of the degree of shared market segments and financial resource allocation on the competitive relationship of two major e-commerce firms: Amazon and Alibaba. Preliminary analysis of financial data and data published by news outlets provide only an incomplete understanding of this competitive relationship. This basic understanding is not enough to critically analyze this associative duality to the satisfaction of academic researchers. Greater insight into this relationship can be provided by processing preliminary data using competitor mapping, an applied statistical theory. Competitor mapping uses the degree of shared market segments and the measurement of certain methods of resource allocation to generate a visual representation of competition.

In order to understand the degree of shared market segments, this study obtains information regarding the size of each firm's worldwide customer base. This is done by scouring large amounts of data to determine the total number of customers served by each firm. From this global number, the number of customers of each firm in a given valued market is determined. In this case, these markets are China and the United States. To guarantee a successful analysis, this derivation process was replicated for each individual firm in each market segment. Essentially, six different sets of data are needed: Alibaba's total number of customers in China, Amazon's total number of customers in China, Alibaba's total number of customers in America, Amazon's total number of customers in America, Alibaba's total number of customers worldwide, and Amazon's total number of customers worldwide. This proved to be a difficult task as this information is not readily available to the public. These numbers provide information about the degree of overlap between the given markets shared by the two firms.

The second part essential for the completion of this statistical study is the measurement of resource allocation that has not previously been defined for the scope of this study. This proved to be another difficult task. The inclusion of a poorly defined resource allocation measure would undermine the validity of the study. Existing research on the airline industry uses similar competitor mapping methods to measure resource allocation by analyzing the type of aircraft in a fleet. In this study, the amount of money spent on essential shipping operations is considered a good indicator. Shipping and logistics are fundamental to any e-commerce business, and therefore, their inclusion in this analysis is appropriate. The acquisition of this dataset was a simpler process, as both firms are publicly traded companies and must therefore divulge all pertinent financial information.

3.2   Theoretical Foundation

Awareness-Motivation-Capability Framework

Figure 1 provides a visualization of the AMC framework for a better interpretation of its elements and the process. The framework incorporates four levels of analysis: (a) the foundations of a competitor, (b) the driving forces of competitive behavior, (c) the implications of behavior on competitive rivalry, and (d) the outcome. Apart from the previously mentioned foundations of competitive interaction in market commonality and resource similarity, the three main drivers of competitive behavior are awareness, motivation, and capability. These drivers affect the likelihood of response and/or attack from competing firms. The increase in the likelihood of a response has implications for a firm's performance and perception of competitors. The feedback loop ensures that there is a repetitive aspect in the framework. This loop is consistent with the perennial nature of competition in business. The loop is present because the resulting impacts on performance and rivalry continually influence a firm's perspective on future competition. For example, action A could decrease a competitor's presence or popularity in a market; this would affect its commonality, and consequently, its motivation to compete, which may alter subsequent strategic perceptions. Thus, the cyclical nature of competition prevails. This framework examines the interactions among competitors and focuses not only on the actions but also on the responses elicited, making it one of the few areas of strategic study that is quintessentially longitudinal (Chen and Miller 2012). Competitive behavior and the resulting reactions are the basis for the creation of competitive advantage and a niche in the market. In the AMC framework, three factors determine the decision-making behavior of a firm: awareness, motivation, and capability. In this paper, it is argued that each aspect of this framework influences not only

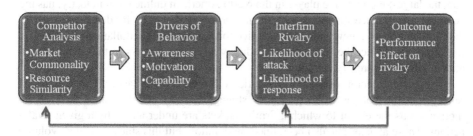

Fig. 1. Competitive dynamic framework

the managers of the firms but also outside stakeholders. We are aware that in addition to the managers of each firm, the stakeholders also exert some influence on organizational direction. The inclusion of stakeholders and their influence on the strategic perceptions of competitive tension is considered here when discussing the macro tensions and perceptions of a company as a whole.

The AMC model has an important perceptual component that acts as the catalyst for the rest of the framework. Perception is the precondition for the rest of the theoretical framework. Logically, it is easy to see how perceptions are incorporated into every level of the framework. First, awareness is synonymous with perception. Second, the motivations of people are formed by their individual perceptions of the world around them. Finally, capability cannot lead to an action unless it is perceived to be adequate, that is, unless a rival, threat, or advantage is perceived by managers or stakeholders as being important enough to warrant the commitment of capability-building resources (Chen and Miller 2012). With this in mind, the entire framework can only imply action by way of these managerial and stakeholder perceptions. Moreover, perceptions and the subsequent motivations and capabilities are variable among each firm.

According to Chen et al. (2007), the first factor that should be considered in this framework is awareness. Awareness is defined as the general cognizance involved in the relationship between a firm and its rival. Generally, in business, there are quite a few factors to be careful about when responding to competition. Many factors and determinants must be collectively considered when making a decision that could directly have dire repercussions in the future, along with financial implications. In the unique industry of e-commerce, this fact is even more pronounced. As mentioned earlier, the e-commerce industry is very dynamic, characterized by rapid development and continual evolution. These characteristics make awareness an important factor for strategy in a robust, yet immature industry. Awareness is also important in relation to the scale of a business's operations. Typically, firms that are well established and easily recognizable in an industry are more likely to initiate substantial strategic attacks on their rivals. They also have a tendency to respond more fiercely and take action to protect their respective market shares (Chen et al. 2007). Smaller players or up-and-coming firms are always aware of industry leaders by virtue of their success and influence on the industry as a whole. Obvious actions taken by a large-scale firm automatically garners the attention of and response by smaller firms because they have significant traction and a reasonable amount of weight. Therefore, the size or scale of a firm is positively correlated with perceived competitive tensions. For example, Amazon, the largest e-commerce player in the North American online retail industry, has the potential to influence the North American market significantly. Its decisions, such as the introduction of a new product or service, or a multi-million dollar acquisition of a rival firm, will affect the nature of the industry and other players in that industry will undeniably be aware of such an action.

The second factor of this framework is the motivation aspect. Motivation is reflected in the volume or frequency of strategic attacks. Chen et al. (2007) define motivation as the extent to which a firm's markets are under attack by a given rival's actions. These attacks can also be defined by duration, but this study considers volume to be a more appropriate indicator. Volume is measured by the total number of strategic attacks highlighted by past instances of competitive action. This historical referencing allows for the measurement of cumulative attack volume. An action, such as entry into new markets or expansion into existing ones, that directly challenges a rival is considered an attack. These initial moves or ensuing counter moves that have immediate implications on current market shares are the most significant types of these attacks. This is especially true when attacks occur in markets that may be of higher value to a firm. For instance, for a firm like Alibaba, the Chinese market is of great importance, whereas a market like Slovakia may not be as important. This is highlighted by the fact that 84% of Alibaba's total revenue comes from within China, while a mere 16% comes from international markets (eCommerceFuel 2014). If an attack is initiated by a competitor to enter the Chinese market and consequently diminish Alibaba's total market share, then an aggressive response is likely to follow. Such actions incentivize internal and external stakeholders to label a firm as a direct competitor based on such attacks and the number of times they occur.

The third contributing factor to the AMC theory is capability, which is the ability of a firm to react to an action taken against it by a competitor. With respect to initial actions taken against a firm's market share, capability is defined by Chen et al. (2007) as a firm's relative resource deployment ability or operational ability to challenge its rival's moves in the marketplace. This operational ability to respond depends on the unique mix of resources a firm has at its disposal. When introducing the idea of resource similarity, each firm, through its development and growth, has a distinct collection of intangible and tangible resources. These resource bundles affect a firm's capabilities and are essential to its operational efficiency. In the event of an attack, these capabilities or resources are deployed by a firm in response to a rival's actions on its market share, with the intention of minimizing it. The firms that have similar or comparable bundles of resources can expect to have similar capabilities. For instance, if two firms have a similar mix of operational resources at their disposal, then an attack by one on another firm with those similar resources is met with a response of similar or equal measure. Firms that have dissimilar or incomparable bundles of resources have divergent strategic capabilities, and therefore, an action by one firm can expect a response that may be unlike the initial action. This paper's idea is consistent with that of previous studies such as Gimeno and Woo (1996) and Chen (1996). Gimeno and Woo (1996) argue that there is a positive correlation between a firm's strategic similarity and its degree of rivalry with competing firms. Chen (1996) showed that the greater the degree of resource similarity between two firms, the greater is the probability that the firms will be involved in competition. Therefore, we can infer that firms with similar capabilities operating in similar markets, assuming that they are cognizant of one another, will experience contentious actions and responses through the perceptions of both internal and external stakeholders.

Figure 2 is a visualization of this AMC framework of competitive tensions described in the prior paragraphs. The awareness, motivation, and capability of firms act as drivers of competitive tension, and their influence on perceptions in turn motivates the frequency of attacks. The focus here is on the AMC framework and its direct impact on the perceptions of firms. Figure 2 is an in-depth and concentrated extension, which outlines a broader understanding of the nature of competition. Each of the main drivers of the AMC framework is defined in terms of their association with real organizational decisions that can be measured by historical events. In this study, the determination of market commonality and resource similarity provides insight into the competitive nature of firms in the e-commerce industry. However, this framework presents an opportunity for a detailed account of AMC.

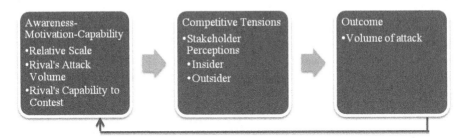

Fig. 2. Competitive tensions

The concepts of market commonality and resource similarity used in the AMC framework provide a basis for measuring the competitive nature of firms in the e-commerce industry. These two constructs reflect rivalry in firms' ability to predict attacks and formulate responses. The comparative and contrasting balance between the two elements highlights the duality of rivalry in the market(s) and the customers served in the market from the perspective of the firm(s).

Competitor Mapping

Competitor mapping is a useful method to define a company's position within a market and in relation to its competitors based on the two previously defined concepts. Researchers have used competitor mapping to investigate the strategic competition of competitor dyads, evaluate actions, and predict reactions (Chen 1996). This method is the focal point of this paper's analysis of competition in the e-commerce industry. Using these data, an illustration of these two firms is constructed in the form of a graph. The concepts of market commonality and resource similarity are used to define the competitive relationship of Alibaba and Amazon.

In existing research, a mathematical formula is used to define commonality in a market based on the number of customers served by competitive firms in a shared market. The formula has been altered for this study to accurately represent the regional markets most important in the e-commerce industry of the United States of America and the People's Republic of China. The study expands this comparison to include more diverse regional markets, such as that of the Asia-Pacific region, North America, and Europe. However, obtaining the data necessary to perform this task proved to be beyond the capacity of this study. Access to data is restricted and requires extensive financial resources. Moreover, the use of greater financial resources does not guarantee the availability of the data required for the study. That said, the United States and the Republic of China are themselves benchmark markets that are satisfactory in this analysis. The development of this formula for market commonality is determined by two factors (Chen 1996): (1) the strategic importance of each of the markets a firm shares with its competitors, and (2) the market share that a competitor manifests with the market. The parameters of the market are defined within each country by the number of people who buy goods online through one of the two retail outlets. The formula is as follows:

$$M_{ab} = {}_{i=1}[(P_{ai}/P_a)(P_{bi}/P_i)] \tag{1}$$

M_{ab} represents the market commonality between **firm A** and **firm B**
P_{ai} represents the total number of customers served by **firm A** in **market i**
P_a represents the total number of customers served by **firm A** in the global market
P_{bi} represents the total number of customers served by **firm B** in **market i**
P_i represents the total number of customers served by all firms in **market i**

Through this formula, this paper compares Alibaba and Amazon in both the Chinese and American markets using data retrieved through various resource portals. For this study, Amazon is the focal firm represented as firm A, and Alibaba is its competitor represented as firm B. Market i will alternate based on the specific calculation and will be represented by the American market population and the Chinese market population for each firm to formulate a comparative dataset of the two most valued e-commerce markets. It is important to compare the regional markets themselves to paint a clear picture of each company's presence and the establishment of a given rival. In order to maintain continuity and facilitate further comparison, this study also calculates the market commonality between these two firms. In the second half of the competitor mapping process, Alibaba will be represented as firm A and Amazon will be represented as firm B.

The second essential part of competitor mapping is resource similarity between the two competing firms. In this case, the resource allocation attributed to the capital expenditure of each firm for shipping and logistical operations provides a reasonable measure of similarity between the two firms. This is because the logistical and shipping arm of a business is a crucial aspect, especially in the e-commerce industry. However, as Sirmon et al. (2008) state, resource management is more important than just resources when rivals' stocks of resources are similar. Therefore, it is a question of not only how much money each firm has, but also how those financial resources are used in their operating strategy. The formula for examining each firm's resource similarity is expressed below:

$$R_{ab} = {}_{i=1}[(P_{am}/P_a)(P_{bm}/P_i)] \tag{2}$$

R_{ab} represents the resource similarity between **firm A** and **firm B**
P_{am} represents the total amount spent on shipping by **firm A** for product delivery
P_a represents the total amount spent on operating expenses by **firm A**
P_{bm} represents the total amount spent on (or in Alibaba's case, invested in) third party logistics by **firm B** for product delivery
P_i represents the total combined amount spent on operating expenses by **firms A** and **firm B**

4 Data Analysis

4.1 Case Study Research

This study investigates the competitive tension between two of the most prominent e-commerce global giants to determine its causes and the degree of intensity. The paper uses case studies to analyze the dynamic nature of the e-commerce industry and incorporate unique aspects of prominent members such as Amazon and Alibaba. In order to do so, the two most influential players are pitted against one another to determine their relationship in the industry. These two companies have regional dominance within their respective countries of origin, despite the market being a global one. Both Amazon and Alibaba want to establish themselves as worldwide frontrunners. Alibaba makes no secret of its global aspirations; however outside of China they face serious competitors, namely Amazon. The same can be said for Amazon in China, which has spent considerable time, money, and effort on expanding its business in China, with little progress to show for it. "Amazon has spent millions of dollars over the last few years in China to build up a series of warehouses and set up other infrastructure to tap the market…but despite that effort the company is still a relatively small player with less than two percent of the market" (Young 2015). Several well-established outlets continue to monitor the actions of both Amazon and Alibaba owing to their importance in the industry. This paper has included the perspectives of respected and reliable resource outlets such as Bloomberg Business, Forbes, The Economist, Investopedia, and Time Magazine.

Within the scope of this competitive setting and the abovementioned theoretical framework, it is essential to include the unbiased opinions of news, research portals, and third-party resources. In addition to these materials, the statements released on behalf of the companies themselves are also valuable. The numerical data provided by the companies are widely considered trustworthy, even though there may be some bias in the information retrieved from other sources. When companies provide statements of purpose, goals, visions, missions, and other forms of rhetoric, there is a tendency to inflate the prospects of the business for creating a positive outlook, even if some things may not be completely truthful. This is why a case study analysis, in addition to an applied frameworks and statistical analysis, is important for this study. As Weisberg (2010) states, a conceptual framework for thinking (about case studies) that extends classical statistical theory is necessary to obtain a deeper understanding. Other research is necessary to critically analyze the data provided in an unbiased manner. Therefore, academic journals and other relevant studies were also used extensively. The objective of this strategy is the accumulation of multi-sourced literature and a rational evaluation of this literature and its association with the e-commerce industry.

Prominent literature explains the importance of case study research. Markus and Lee (1999) state that the goal of case study research is to promote intensive research through empirical studies, which can serve as models of how to do intensive research and will illustrate the criteria through which such research can be evaluated. There is also a degree of diversity involved in the use of case study research that produces a greater variety and range of learning. Unsurprisingly, case study analysis is used frequently as a teaching method in business schools, as real life experiences are invaluable commodities. Practicality and pragmatism are two aspects of case study research that make it valuable. Real life examples from businesses provide a different, if not more supportive, way of understanding cause and effect. This paper uses case study analysis for highlighting the differences or similarities between two major businesses within this unique and dynamic industry. This study explores the varying degrees of duality in the two fundamental aspects of competitive tension, and explains the way in which the industry works.

Case study analysis has fewer limitations than other methods such as laboratory testing or heavily concentrated statistical theory, as these methods use a plethora of control, dependent, and independent variables to deduce conclusions. This can be valuable in focused quantitative analysis, but in more qualitative research, these methods are less effective. Through case study research, researchers find that natural observations are more likely to produce a clearer understanding of the practical occurrences; that is, a simple observation of the scenario without the application of outside forces (controls) produces interesting results. Lee (1989) states that a case researcher must actively apply his or her ingenuity in order to make predictions that take advantage of natural controls and treatments that are either already in place or are likely to occur. He argues that case studies can be conducted as a form of a natural experiment, which in itself is already a conventional form of research (Lee 1989). Case studies are widely used in academic research and teaching. Therefore, they are used in this paper to investigate the competitive relationship between Alibaba and Amazon in the e-commerce industry.

4.2 Analysis of Competitive Dynamics (Awareness-Motivation-Capability)

The AMC framework, as it pertains to the perceived competitive tensions between the two companies, is an important first step in understanding where these two companies stand in relation to one another (not their physical location, but their relationship in the realm of markets and resources). These data are simply an observation of reality based on previously studied theory. This theory will paint a picture that allows readers to gain insight into the relationship of Alibaba and Amazon in terms of their market commonality and resource similarity. This paper uses data regarding the numbers of customers in valuable regional markets and the allocation of capital resources to shipping and logistics. The relationship between market commonality and resource similarity is not universal and can be expressed in many different ways. These two companies are rich in strategy and diverse in nature, which provides many different avenues for critique and comparison. Every single action and reaction can be scrutinized and discussed. However, the tensions or relationship between these two firms is largely unknown. What we perceive is superficial in that society can only really know what these powerful companies choose to reveal. How one company feels about the other will never truly be known. Therefore, in this paper, we speculate the nature of the relationship between the two firms using competitive dynamics theory and the AMC framework.

This observational strategy also recognizes that despite the competitive nature of these firms, there are several opportunities for a mutually beneficial relationship. This will be analyzed later. In this context, coopetition is seen in Amazon's decision to open up its own store on Alibaba's Tmall website. This is because Amazon is continually attempting to expand into the Chinese market of more than one billion people. In terms of awareness, this paper recognizes that the pair of firms has an abundance of information at their disposal. This position would not be possible if not for the intelligent collection of people and other miscellaneous resources collected over the years, which are now at their disposal. We assume that both Alibaba and Amazon are aware of what the other firm is doing, in both their respective markets and those worldwide. This has to do with the development of technology that has spurred a technological revolution, the widespread use of the Internet that shows no signs of decline, and the seemingly unquantifiable amount of information created by the combination of these two factors. These businesses are equipped with capabilities to collect massive amounts of data on individuals and other companies. Facebook collects data on things such as where you are going, what you are doing, who your friends are, and what your friends are doing. Google collects data on things such as your location, search history, email content, and what you watch. The same can be said for data collection on competitors or rivals. The fact that these companies collect unquantifiable amount of data suggests that what people or civilian consumers are doing is as important to businesses as what their competitors are doing.

In terms of motivation, we must assume that both companies are highly motivated in their pursuit of continually expanding their market share in the global e-commerce industry. Despite what the data say about their relationship and perceived tensions, both Amazon and Alibaba are committed to expanding the scale of their respective businesses. The continual growth for any business must incorporate the movement and entry into new markets. Once limitations are encountered in a given region, a business will move on to other markets to continue to expand and satisfy shareholder expectations. This also reverts to the theme of the provision of Internet-enabled expansion opportunities for businesses that are not limited by geography. Motivation, as defined by this framework, can be measured by the number of attacks or instances of competitors encroaching on shared markets. These data suggest that Alibaba's presence in the US market is small, and the same can be said for Amazon's presence in China. However, despite their current imposition, the effort to increase that imposition is continual regardless of whether it yields positive outcomes.

Finally, the capability of a firm is a direct reflection of its ability to react to actions taken by rivals that are directed against it. This can manifest itself in a number of different ways. Most decision-making actions may not be a direct attack on a given rival, but rather an indirect method of doing good business. Business, and more specifically the battle for market share, is marred in attrition. Businesses are constantly engaged in creating new ways to bring in as many people as they can to participate in their business in comparison to other firms. In the case of Alibaba and Amazon, both firms have sufficient capability to react to the strategic actions of the other. The reality is that the majority of successful businesses have rigorous hierarchical structures, whose strategic actions are subject to substantial critical analysis by decision-making bodies before being implemented. Numerous people and departments must sign off on any action to limit the number of ineffective decisions made. Firms that control as much capital as Alibaba and Amazon do can physically perform several tasks, but their business environment may not allow all of them to be performed, as some accountability needs to be enforced. Reputation is extremely important, and businesses do everything they can to create a valuable one. The resource bundles of Amazon and Alibaba are pretty similar in that they both have significant advertising services, are involved in company-specific technology platforms, match buyers and sellers, have individual online storefronts, and operate a membership-fee-structured revenue stream (Zucchi 2015). Even though these firms differ in several respects, their general composition is similar enough for this paper to state confidently that their capabilities are equally widespread.

4.3 Analysis of Competitor Mapping

Competitor mapping is the process of pitting two elemental concepts against one another in order to analyze the competitive bond between two firms in a shared industry. In this paper, Alibaba and Amazon are the two focal firms of analysis. The relationship these two companies share in terms of the two foundational concepts of resource similarity and market commonality are broken down into four quadrants. Each quadrant represents a fundamental relationship of the competitor mapping process. Furthermore, each quadrant denotes a different comparative relationship combination linked to the degree of each concept (market commonality and resource similarity). The quadrants have different implications on the competitive tensions between the pair of firms. These combination relationships are an extension of the natural observations of competitive dynamics. Figure 3 illustrates the case of low market commonality and high resource similarity.

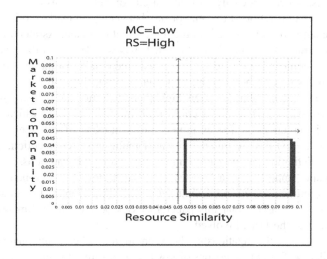

Fig. 3. High resource similarity vs low market commonality - *Quadrant 1*

Figure 3 shows two firms that have similar or comparable bundles of organizational resources, but are only currently competing in a few similar markets. This paper formulates hypotheses based on these concepts and their meanings. If two firms have similar resources but operate in dissimilar markets, then strategic actions will be taken to encroach on one another's markets to improve global reach and increase the potential for customers. Despite this push or encroachment, they are not yet labeled as competitors because of their dissimilar market composition. This assumption is based on a fundamental principal of business that growth and customer acquisition are important goals for any business entity. The initial action or the initial attacker can also expect to be met with a similar response or retaliatory action, because regardless of the resources one may choose to allocate towards the success of this initial decision, the other firm will respond with the same volume or intensity in terms of resources. Simply put, decisions will be made to enter new markets, and any action taken is reciprocated by way of the action and response dyad of competition. This is the theoretical association between high resource similarity and low market commonality.

The relationship between high market commonality and high resource similarity may yield the most competitive relationship in the entirety of this analysis. Firms that share numerous markets and have comparatively similar bundles of resources are categorized as fierce competitors. They frequently engage on a multimarket level, which typically generates large numbers of actions and responses. Chen (1996) states that two firms will experience great tension if they compete directly in many markets, and more fundamentally, if each is a key player in markets vital to the other. This has significant implications in the present case study because both Amazon and Alibaba have considerable regional dominance. Alibaba is dominating China, and Amazon views the Chinese market as a high priority in its quest for global expansion. Similarly, Amazon has a stranglehold on the American market, which is an area that Alibaba is making significant strides in trying to cultivate. Therefore, each firm has something that the other wants, which is a catalyst for increased competitive tension.

Figure 4 is an illustration of the relationship between high market commonality and high resource similarity represented in the upper right quadrant. As mentioned before, high resource similarity means that a firm that makes a decision to act can expect a response that is comparable owing to similar resource bundles of the two firms. This is the theoretical association between high market commonality and high resource similarity. Next, the relationship between high market commonality and low resource similarity is considered. Here, firms compete in multiple markets, which means that they engage in competition for customers on a number of different levels. It is important to note that despite this definition of multimarket competition, the characteristics of the definition can be narrow in its derivation. We cannot assume that each market is equally important to a firm. There is an asymmetrical property associated with market commonality. For now, the narrow definition of market commonality will suffice because the markets that this paper focusses on are equally important as the other in relation to the firms involved.

Figure 5 presents the relationship between high market commonality and low resource similarity. The characteristics of low resource similarity are related to a firm's ability to implement a strategy as compared to its competitors. This paper has already outlined the implications of similar resource bundles and its effect on the competitive relationships of firms. To review, firms with similar resource bundles are likely to have similar strategic capabilities and competitive vulnerabilities in the marketplace. Similarly, firms with divergent resource bundles are likely to have diverse competitive

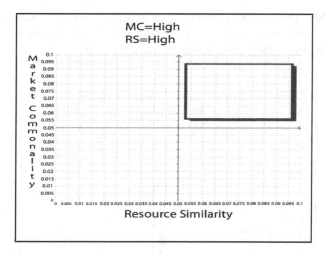

Fig. 4. High resource similarity vs high market commonality - *Quadrant 2*

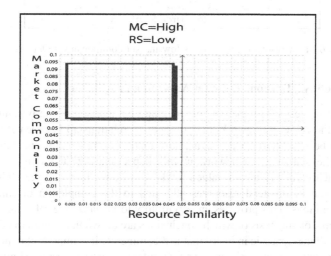

Fig. 5. Low resource similarity vs high market commonality - *Quadrant 3*

repertoires to draw on because of the unique profiles of their strategic resources (Chen 1996). Therefore, despite the representation that firm relationships located in the upper left quadrant of this grid compete in similar markets, their resource bundles are dissimilar, which implies that their strategic capabilities are dissimilar. This means that competitive actions are frequent; however, the responses they elicit are unique and representative of the individual firms.

Finally, the last quadrant of the grid illustrates low market commonality and low resource similarity. The pairs of firms that do not have a high degree of market commonality, have dissimilar allocation of resources, and have regional dominance in opposing highly valued markets have a highly dynamic competitive relationship. This relationship is representative of that of Amazon and Alibaba. Figure 6 presents this relationship located in the bottom left quadrant of the grid.

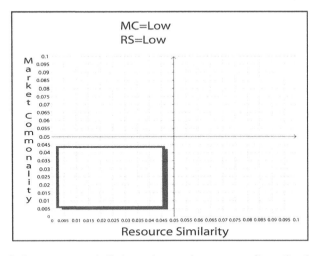

Fig. 6. Low resource similarity vs low market commonality - *Quadrant 4*

4.4   Competitive Tensions Between Amazon and Alibaba

Based on the process of competitor mapping and comparison of the two fundamental concepts of the AMC framework, the competitive relationship between Amazon and Alibaba can be characterized as having low market commonality and low resource similarity. The statistical framework in this paper analyzes the number of shared customers each company has in valued markets in both China and the United States, and the amount of financial resources allocated to the completion of shipping and logistical operations. This provides an opportunity to visualize the relational duality of these two firms, on the basis of which conclusions are drawn in addition to other factors associated with the dynamic industry that is e-commerce. It is important to note that the results of the statistical analysis produced an outcome that was highly similar for both firms. Each relationship was located in the same quadrant as the other, which suggests a parallel situational relationship. Here, industry position refers to the current composition of influential companies in e-commerce and their long-term strategic goals pertaining to the markets they currently dominate and the valued markets they are taking competitive actions towards entering.

When using Amazon as the focal firm, Alibaba is located in *quadrant 4*. When using Alibaba as the focal firm, Amazon is also located in *quadrant 4*. This likeness of the plotted location is because of the similarities in the current position of each firm in the industry. The current industry position for each firm can be explained by the symmetry associated with a strong presence in each firm's respective home market, and a comparatively weak presence in the opposing valued market. Amazon has a commanding influence in the American market and spends tens of billions of dollars on operational expenses, while Alibaba operates on a small scale in terms of market share and financial investment. Conversely, Alibaba has an established monopoly and spends a majority of its operational expenses within the Chinese market, while Amazon is insignificant and spends only a small percentage of its overall expenses there. Thus, the two firms are similar in that they are small players in the market of their respective counterparts while dominating their home markets. This similarity in market position is revealed in the borderline symmetrical position they hold in the plotting of the grid during the competitor mapping process. The position of each firm's relationship in relation to the other is an important indicator for how perceived tensions can be analyzed.

Using Amazon as the focal firm in the American market, Fig. 7 represents the competitive relationship. With respect to Amazon in the United States, Alibaba has a dissimilar resource allocation in that it does not have the substantial logistical prowess in the United States that Amazon does. Alibaba's minimal traction in the United States comes in the form of a 2013 investment in the American e-commerce and logistical start-up ShopRunner, worth 202 million USD (Werkun et al. 2014). This miniscule investment is pale in comparison to the billions spent on shipping by Amazon each year, which explains the dissimilarity of resource allocation seen here.

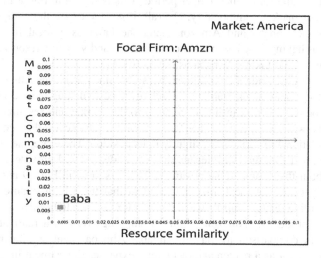

Fig. 7. Amazon vs alibaba – competitive relationship in the American market

Using Amazon as the focal firm, a similar picture emerges in the Chinese market, as shown in Fig. 8. Market commonality is slightly higher in China because of the intensity of Amazon's efforts to expand into China. Despite these efforts, Amazon has not yet garnered any real market share.

Amazon has recently opened a storefront on Alibaba's Tmall. It also operates Amazon.cn, which was formerly a Chinese e-retailer called Joyo.com. Amazon acquired it in 2004 as one of many strategic actions aimed at expanding into the

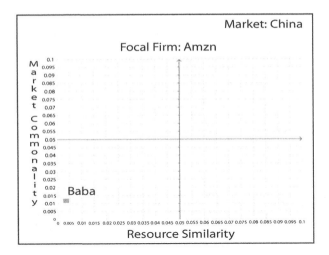

Fig. 8. Amazon vs alibaba – competitive relationship in the Chinese market

Chinese market (Wiles 2004). Compared to Alibaba, which has only recently initiated strategic expansive actions in the American market, Amazon has been working towards expanding into China for a much longer period. This is shown in Fig. 8 in the slightly higher degree of market commonality. Overall, it is easy to see the competitive relationship between Alibaba and Amazon using the latter as a focal firm; it can be characterized as having very low market commonality and very low resource similarity. The same can be said taking Alibaba as the focal firm.

In an attempt to provide a thorough analysis, this study also uses Alibaba as a focal firm in order to present an alternative perspective, since switching the focal firms of the study may produce different results in relation to the market and resource constructs. However, this was not the case. Irrespective of the choice of the focal firm, the competitive relationship varied only slightly in its position. Therefore, the change in the choice of the focal firm did not offer alternative results, but did provide an interesting conclusion about their comparable positions. This is shown in Figs. 9 and 10.

Using Alibaba as the focal firm, it is evident that there is little difference in the overall competitive relationship between the two firms. The slight increase in resource similarity is a product of Amazon's inflated shipping costs as a ratio of its overall expense cost in comparison to that of Alibaba. Alibaba spent significantly less on shipping in total, but as a percentage of its total expenses, investment in shipping and logistics accounts for nearly a quarter of Alibaba's overall costs (1207 million/4470 million). Amazon's shipping costs are about four times that of Alibaba, and accounts for roughly a twentieth of its overall expenses (4223 million/88810 million).

The conclusion of this methodological approach to competitive dynamics yields some interesting results that would not otherwise be obtained. According to competitor mapping, Amazon and Alibaba are not yet identified as major competitors by virtue of their opposing market contact and dissimilar allocation of financial resources. According to competitor mapping, these firms are considered fierce competitors when the duality of their relationship is located in *quadrant 2*. This indicates high resource

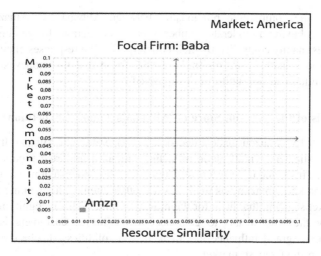

Fig. 9. Alibaba vs Amazon – competitive relationship in the American market

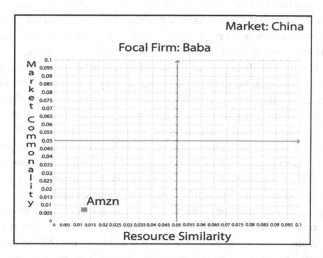

Fig. 10. Alibaba vs Amazon – competitive relationship in the Chinese market

similarity and high market commonality. Regardless of this classification, Alibaba and Amazon do, in theory and in practice, engage on a competitive level based on their lack of market commonality. Chen (1996), Gimeno and Woo (1996), and Baum and Korn (1999) affirm that:

"Firms with more multimarket contact (high market commonality) tend to take more conservative stances toward each other, leading to less rivalry, as reflected by high market prices. An increase in multimarket contact reduces rivalries rates of entry into each other's markets thus minimizing this particularly aggressive competitive incursion. Thus, with respect to the initiation of attacks, it appears that an attacker will be less likely to target rivals with high market commonality than those with low market commonality, because the stake involved is very substantial" (Chen 1996).

Therefore, the competitive relationship between Alibaba and Amazon can be characterized as having a moderate number of attacks or actions for gaining entry into valued markets mainly controlled by their counterparts. The responses elicited by these actions into those markets are dissimilar in the sense that the money spent on the response is unique and non-uniform, as expected.

4.5   Analysis of the Amazon Store on Alibaba's Tmall (Coopetition)

On March 5, 2015, Amazon made a decision to cooperate with Alibaba. It launched a storefront on Alibaba's Tmall platform selling premium imported products from the United States. This is the most recent strategic move by Amazon aimed at tapping into the Chinese market. Amazon has made several other attempts at cracking the difficult Chinese market segment that have for the most part failed. This recognition of Alibaba as potentially the only means to gain access to Chinese customers is an honest recognition of its grand influence in China. It also solidifies Alibaba's monopolistic stranglehold on the Chinese market.

In making this decision to join Tmall, Amazon must adhere to all the fee and profit-sharing structures associated with being a Tmall merchant. That is to say, no matter how well Amazon does on Tmall, a portion of all that profit will go to Alibaba. Therefore, Alibaba is now in an even greater position of power. Amazon has conceded some of its efforts in recognition that it can only do well if it associates itself with the dominant market player. Basically, at some point you need to go where the fish are (Soper 2015). In hindsight, these strategic decisions may have been imminent for some time. Amazon has consistently poured substantial resources and efforts into the expansion of its business into the Chinese marketplace in an attempt to steal a share of the growing number of online customers in China. Dismal results show that these efforts have not been successful, as its market share has consistently declined over the past few years. These past decisions have seen little return on investment, as Alibaba's domination in China continues to grow. This most recent strategic action by Amazon is interesting and provides some linkage between the outcomes of this study's competitor mapping analysis and coopetition, which is related to the AMC framework.

Coopetition can yield beneficial results for both firms involved by way of shared resource allocation (Peng et al. 2012). Amazon's ability to use Alibaba's online platform and established brand to ameliorate their diminishing returns and other failed strategies is a perfect example of this. Amazon uses some of Alibaba's resources, and in exchange, Alibaba profits from the situation. There is also congruence between the results of this study in terms of Alibaba and Amazon's competitive relationship. Direct competitors whose competitive relationship is characterized by high resource similarity and high market commonality do not engage in coopetition due to the fierce nature of their relationship. Therefore, Alibaba and Amazon are not considered direct competitors (their relationship is characterized by low resource similarity and low market commonality); therefore, this presents an opportunity to analyze coopetition in this case. Amazon decided to act when this opportunity to cooperate and share resources presented itself. In this case, resources include Amazon's ability to use a popular existing platform to improve the popularity of its brand, increase its reach to customers who may not have been otherwise aware of its brand, and increase sales in a market that has historically been difficult to penetrate. Therefore, in light of the nature of Amazon and Alibaba's positional relationship within the e-commerce industry, and the ability for coopetition to yield mutually beneficial results, Amazon's participation on Tmall is a perfect illustration of the linkage and validity of competitive dynamics theory and the AMC framework.

5 Conclusion

This study is a representation of the current progress of technology and innovation in the realm of business. The technological revolution has spurred new kinds of industries that use the Internet and various connected platforms to create value and wealth in places that would otherwise have been unrealized. The combination of massive amounts of data creation and the use of faster, smarter, and smaller technologies represent the forefront of innovation and subsequently the evolution of modern day human interaction. Engrained within this dynamic space are a handful of companies that are leading in their own respective way, carving out a niche for themselves in a lucrative global marketplace rampant with consumerism and an unquenchable thirst for fresh ideas. This study focuses on the e-commerce industry in terms of two giant companies (Amazon and Alibaba). The Internet itself is only about a half-century old, and the majority of its development has occurred in the past two decades. There is still much to learn about the nature of its use and the direction of its evolution.

The study investigates the phenomenon of the Internet and its resulting impact on the world of business. The Internet has produced companies with unlimited reach, astronomical frequency of use, and massive valuations through growth potential and scalability. Therefore, the value of these companies is much greater than has been hitherto seen. In this context, the present study investigates the impact of technological development on society, the resulting emergence of a new industry, the characteristics of that industry, and the nature of competition highlighted by its two most prominent firms. Amazon and Alibaba are the most innovative and popular companies in the industry today.

Competitive dynamics theory was used to explore the nature of competition. This theory encompasses a wide variety of possible research directions from macro to micro factors. The study analyzes the AMC factors associated with each individual firm. This AMC framework describes a given firm's propensity to take strategic or responsive actions against its competitors. The propensity to act is highlighted by competitive tensions or the perceptions a firm may have in relation to its competitor's position in the same industry—this is the competitive relationship. The fundamental principle for deriving this relationship associated with AMC is presented through a statistical method known as competitor mapping. Competitor mapping associates commonality in market structure with similarity in organizational resource allocation to derive a relationship that can be plotted as a location on a grid. This method was used considering the simplicity of interpreting its results

The preliminary results of this study identify the characteristics of all groupings of the relationship between the fundamental constructs of the AMC theory—resource similarity and market commonality. There are four distinct possible relationships between market commonality and resource similarity: low-low, low-high, high-low, and high-high. Each possible outcome relationship is associated with a given set of competitive characteristics, which this study determines as the likelihood or volume of attacks and the types of responses such attacks would elicit. For example, the likelihood of an attack(s) from firms that have a low-high relationship differs from that for a pair of firms that have a high-low relationship. Furthermore, the response elicited from a pair of firms that have a low-low relationship is also different from that for firms that have a high-high relationship. Essentially, each relationship grouping has different implications for competitive nature and strategic decision-making.

The study found that companies with low market commonality and low resource similarity are not direct competitors, but compete nonetheless. Companies that have high resource similarity and high market commonality are categorized as direct competitors according to Chen (1996). Firms with low market commonality and low resource similarity can be categorized as moderate competitors. Alibaba and Amazon fall into this category. The moderately competitive nature of these two firms means that there is an underlying recognition of competitive tension, even though there is an opportunity to cooperate.

The results directly reflect the competitive relationship of these two firms in reality. Amazon's recent decision to collaborate with Alibaba in China validates the theoretical and statistical outcome of the applied methodology and theory. The results of this study also suggest that coopetition has the potential for mutually beneficial outcomes. Since coopetition is a recent development in this relationship, this study cannot conclusively state that Amazon's decision to join Alibaba's Tmall platform will be successful. Perhaps, this point can be revisited in order to analyze the impact of this decision on both firms in the future.

References

Alba, D.: 50 Years On, Moore's Law Still Pushes Tech to Double Down, 19 April 2015. http://www.wired.com/2015/04/50-years-moores-law-still-pushes-tech-double/. Accessed 22 Apr 2015

Baum, J.A.C., Korn, H.J.: Dynamics of dyadic competitive interaction. Strateg. Manag. J. 20(3), 251–278 (1999)

Chen, M.-J.: Competitor analysis and interfirm rivalry: toward a theoretical integration. Acad. Manag. Rev. 21(1), 100–134 (1996)

Chen, M.-J.: Reconceptualizing the competition-cooperation relationship. J. Manag. Inquiry 17(4), 288–304 (2008)

Chen, M.-J., Miller, D.: Competitive dynamics: themes, trends, and a prospective research platform. Acad. Manag. Ann. 6(1), 135–210 (2012)

Chen, M.-J., Su, K.-H., Tsai, W.: Competitive tensions: the awareness-motivation-capability perspective. Acad. Manag. J. 50(1), 101–118 (2007)

DeSarbo, W.S., Grewal, R., Wind, J.: Who competes with whom? A demand-based perspective for identifying and representing asymmetric competition. Strateg. Manag. J. 27(2), 101–129 (2006)

eCommerceFuel: Alibaba vs. Amazon: An In-Depth Comparison of Two eCommerce Giants, 24 October 2014. http://www.ecommercefuel.com/alibaba-vs-amazon/. Accessed 18 May 2015

Gimeno, J., Woo, C.Y.: Hypercompetition in a multimarket environment: the role of strategic similarity and multimarket contact in competitive de-escalation. Inst. Oper. Res. Manag. Sci. 7(3), 322–341 (1996)

Gnyawali, D.R., He, J., Madhavan, R.: Impact of co-opetition on firm competitive behavior: an empirical examination. J. Manag. 32(4), 507–530 (2006)

Haleblian, J., McNamara, G., Kolev, K., Dykes, B.J.: Exploring firm characteristics that differentiate leaders from followers in industry merger waves: a competitive dynamics perspective. Strateg. Manag. J. 33, 1037–1052 (2012)

Hunter, M.: A short history of business and entrepreneurial evolution during the 20th century: trends for the new millennium. Geopolit. Hist. Int. Relat. 5(1), 44–98 (2013)

Laudon, K.C.: E-Commerce 2015, 11th edn. Prentice Hall, Upper Saddle River (2014)

Lee, A.S.: A scientific methodology for MIS case studies. Manag. Inf. Syst. Q. 33–50 (1989)

Markus, M.L., Lee, A.S.: Special issue on intensive research in information systems: using qualitative, interpretive, and case study methods to study information technology - foreward. Manag. Inf. Syst. Q. 23(1), 35–38 (1999)

OED Online: The Oxford English Dictionary. Oxford University Press (2015)

Peng, T.-J.A., Pike, S., Yang, J.C.-H., Roos, G.: Is cooperation with competitors a good idea? An example in practice. Br. J. Manag. 23, 523–560 (2012)

Porter, M.E.: Strategy and the internet. Harvard Bus. Rev. 62–78 (2001)

Sirmon, D.G., Gove, S., Hitt, M.A.: Resource management in dyadic competitive rivalry: the effects of resource bundling and deployment. Acad. Manag. J. 51(5), 919–935 (2008)

Soper, S.: Amazon Opens Store on Alibaba's Tmall for Chinese Shoppers, 6 March 2015. http://www.bloomberg.com/news/articles/2015-03-05/amazon-opens-store-on-alibaba-s-tmall-to-reach-chinese-shoppers. Accessed 8 June 2015

Tsai, W.: Social structure of coopetition within a multiunit organization: coordination, competition, and intraorganizational knowledge sharing. Organ. Sci. 13(2), 179–190 (2002)

Weisberg, H.I.: Bias and Causation: Models and Judgement for Valid Comparisons (2010)

Werkun, E., Sheridan, E.J., Xu, A.: Alibaba Group Holdings Limited: A global eCommerce Leader Emerges. UBS Global Research, 14 October 2014

Wiles, G.: Amazon.com to Acquire China's Joyo.com for $75 Mln. Bloomberg, 19 August 2004. http://www.bloomberg.com/apps/news?pid=newsarchive&sid=aTvLoEuMIazQ

Young, D.: Amazon Retreating In China? Not Exactly, 9 March 2015. http://www.forbes.com/sites/dougyoung/2015/03/09/amazon-retreating-in-china-not-exactly/. Accessed 13 May 2015

Zucchi, K.: Navigating E-commerce: Alibaba, eBay and Amazon (2015). http://www.investopedia.com/articles/investing/102814/navigating-ecommerce-alibaba-ebay-and-amazon.asp. Accessed 23 May 2015

Temporal Ownership Boundary in Sharing Economy

Huihui Chi[1], Wei Zhou[1,2(✉)], and Selwyn Piramuthu[3]

[1] Information and Operations Management, ESCP Europe, Paris, France
huihui.chi@edu.escpeurope.eu, wzhou@escpeurope.eu
[2] Big Data and Business Analytics, ESCP Europe, Paris, France
[3] Information Systems and Operations Management, University of Florida,
Gainesville, FL, USA
selwyn.piramuthu@warrington.ufl.edu

Abstract. We investigate the temporal ownership boundary that exists in the sharing economy. We find that temporal factors play an important role in the decisions of collaborative contribution. A collaborative contributor need not only consider the engagement duration and the potential income, but also the holding/inventory/maintenance costs during its ownership. We define the temporal ownership boundary as the limit when the owner is indifferent of transferring the ownership from its current in-usage or sharing status. By this definition, we can decompose a merchandise as two substitute goods: the ownership good and the transferring good. The ownership good can be consumed or shared by the owner. The transferring good can either be given as a gift or be resold for an income. The temporal ownership boundary can be found by considering the owner's holding cost, various transaction costs, and the potential income from the sharing economy activities. We find that there exists various conditions when this boundary may lean towards sharing, gift giving or reselling.

Keywords: Temporal ownership boundary · Sharing economy · Resale market · Gift economy

1 Introduction

In recent years, many sharing marketplaces have emerged, targeting various economic segments, for example AirBnB and Roomorama for lodging, SnapGoods for tools, RelayRides for cars, Wheelz for bikes, Uber and Lyft for ad hoc taxi services, etc. Services can also be shared, for example peer-to-peer lending, crowdfunding, couch-surfing, coworking, knowledge and talent-sharing. We foresee that sharing economy will keep evolving and become more flexible and eminent in the near future because of its fundamental economic drivers.

In general, the sharing economy refers to the economic activities when goods or services are arranged to be contributed and shared among a group of consumers. It is normally characterized by a discounted price and a partial income for the collaborative contributors. Before the end of life (EoL) of a good (or a service subscription), the

© Springer International Publishing AG 2017
M. Fan et al. (Eds.): WeB 2016, LNBIP 296, pp. 54–66, 2017.
https://doi.org/10.1007/978-3-319-69644-7_5

owner can freely use, share, give, or sell the good's remaining value. For example, old clothes, musical instruments, books, cars, apartments, their value can be realized by above-mentioned economic forms. What affects an owner's decision to choose one from the others?

To answer this question, we must consider the value of the goods, its inventory and holding cost, the transaction costs, and potential income from the sales, usage, or collaborative contribution. Most of these parameters are temporal factors. For instance, a costly inventory would normally give the owner a strong incentive to transfer its ownership by resale or gift-giving. On the contrary, a high remaining value of the good would make the owner willing to keep the ownership for individual or group consumption. We are motivated to investigate the properties of these temporal ownership variables.

A good's remaining ownership can be transferred. It creates other two emerging markets: the reselling market and the gift economy. The gift economy refers to the economic activities that aim to transfer goods or services freely to other individuals without an agreed method of quid pro quo. The economy of reselling represents the transfer of the remaining value of goods accompanied by a resale price. The goods that are idle for one person may be needed by others, so they can make a deal by sharing, giving, or reselling. During the ownership period, we make a simplified assumption by considering individual consumption as a special case of collective consumption when only the owner utilizes the good/service. By considering all the factors that influence the decision making, with general rationality, the owner is more willing to keep the ownership when the future income and the value exceed the costs of holding it. When the holding cost is significantly high, the owner would be more likely to transfer the ownership. There should exist an equilibrium where the owner is indifferent to sharing, giving, or reselling. The indifference point is further adjustable by the sharing network, charity organizations, the taxation policy, and by resale marketplace.

We aim to investigate the temporal ownership boundary that exists in the sharing economy. We study the temporal factors including the inventory holding cost, the potential collaborative income, and individual utility from consumption with various time stamps. We define the temporal ownership boundary as the limit when the owner is indifferent to transferring the ownership from its current in-usage or sharing status. We base our analysis on two variations of substitute modelling and consider the properties of social welfare by incorporating the utility functions of different players. We find that there exist various conditions when this boundary may lean towards sharing, gift giving or reselling. We show that both individual utility and total social welfare can be optimized by adjusting the incentives, the transaction costs, and eventually the time of ownership transfer of the goods. Our results bring meaningful and interesting insights to today's sharing, gift, and resale platform companies on how to improve the efficiency and competitiveness.

The remainder of this paper is organized as follows. In Sect. 2, we provide a brief review of literature in today's sharing economy, resale marketplace and gift economy. In Sect. 3 we propose a substitute model that defines the temporal ownership boundary. We discuss the results and draw managerial implications in Sect. 4. We make concluding remarks and give guidance to future research in Sect. 5.

2   Literature Review

From the existing literature in sharing economy, gift economy and resale market, we find a common agreement that all forms of ownership must create real consumer value at the end. The concept of sharing bikes (Wheelz), cars (Uber), or houses (Airbnb) begins to become more and more popular [5]. In order to obtain the stable mobility, existing shared mobility business models try hard to find the optimal relationship between good owners and receivers. What's more, sharing economy now achieves success in the competition with concrete firms and makes itself different to acquire market share [13]. The economy of sharing is often linked to the collaborative consumption [2]. In terms of how we think about ownership, collaborative consumption is often considered as important as the Industrial Revolution. Almost all industries are involved in this ongoing disruptive change of sharing economy and collaborative consumption. People can use collaborative consumption as a force to effect the sustainable development and a method to strengthen communities [3]. On top of that, owing to information and communications technologies, collaborative consumption develops rapidly [7]. Moreover, different factors like sustainability play important roles in motivating the participation in CC.

However, its dark side needs to be deal with when the sharing economy grows up [8], which means to gain unfair advantages like regulatory arbitrage should be avoided. And democratizing the ownership and governance of the platform would help to control the power of new technologies [11]. On the other hand, there are still fleets and inventory costs even in the sharing economy [12]. New sharing economy market models like re-engineered consumption models are needed.

Supposing if there is no inventory cost, the owner of the goods/services would have the intention to keep them with any residual value. If there exists a reward to transfer the ownership as a gift or certain holding cost, the tendency to keep the goods may withdraw. It creates the economy of gifts. With non-zero inventory cost, depreciating value of the good and taxation benefit, the owner might make a negative utility if he/she holds the good. The gift economy, however, is not always attractive [9], and it can push people away and seek the valorised market as an alternative option. What's more, people pay little attention and hardly show their understanding of gift giving [4], because of the privacy and conceptual framework of this activity.

Resale market heavily depreciate machines produced in these dispute-affected equipment [10]. Moreover, components of these machines are resold more frequently and receive lower list prices. In the standard auction, there is a bidder in resale market who doesn't have any use value for the good on sale [6]. When resale leads the auction, there is an equilibrium in the auction-plus-resale game, which would determine the bidding price [1]. But in a perfect resale market, the auction with resale would not be the best choice for the seller.

3   Research Methodology

3.1   Temporal Boundary and Ownership Substitution

Today, the barrier of group trade, collaborative consumption, and donation has greatly diminished compared to the recent past. It creates new yet phenomenally large business communities to share, resell and gift the remaining value of goods and services. We observe certain similarities among the mentioned three forms by considering the time when the good's ownership is transferred. Individual and collaborative consumption is characterized by holding the ownership. Collaborative consumption does not strictly follow the ownership if the consumer only "rents". Because the focus of this research is on the boundary of sharing and gift, we emphasize on the good owner's decision in the following model development. Gift and resale are similar because in both forms the ownership will be transferred. Figure 1 illustrates how we can decompose a good or service simply based on its expected life and the time point of ownership transfer.

Thus, a good can be decomposed into two substitutes by specifying the temporal boundary when the ownership is transferred, ranging from time zero to its end of life (Fig. 1). Let T represent the end of life time stamp of a good. We decompose the good/service in two parts: the ownership part (P_s) and the detachment part (P_g or P_r). P_g represents the remaining part to be given as a gift and P_r indicates that the remaining value will be resold on a reselling market. Whenever the good still belongs to the owner, he/she has to choose whether to separate the ownership in the future, when (t), and how (gift or resale).

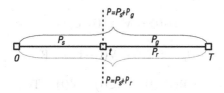

Fig. 1. Decomposition of a good

3.2 Analysis of the Optimal Decision

The gift economy is different from a free supply-demand market because the goods are not well organized, listed and marketed to the consumers. It involves transaction costs for both donors and receivers to give and find the right goods. In this sense, the intermediary plays a very important role in reducing the transaction costs from both side by giving effort to improve the service.

We consider the intrinsic value of goods, inventory holding cost, transaction cost, and good-will rewards in the gift economy model.

We use the following list of notations:

V: Value of the good
HC: Holding cost of the good
R(v):   Good-will reward of giving a gift
S: Income from sharing
TC_1:   Transaction cost of sharing
TC_2:   Transaction cost of gift
TC_3:   Transaction cost of resale
T: Estimated remaining life of the good from time zero

Re(v):   Resale price
i: Time interest/discount rate
U_r: Utility from resale
U_g: Utility from gift giving
U_1:   Utility from owning the good
U_2:   Utility from detaching the good
σt: Sharing income volatility

At any time point of a good before its end of life, the owner has three choices: (1) to give the good as a gift, (2) to share(use) it, (3) to resell it. The utility of sharing/using the good is the value of the good (V) plus the income from sharing/using it (S) minus the holding/maintenance cost (HC) and the transaction cost (TC_1) as represented in (1). The utility of gift giving consists of the inventory holding credit (HC) plus the reward (R) minus the value of the good (V) and the transaction cost(TC_2), Eq. (2). The utility of reselling the good is the price of the good according to the value of the good (V) minus the holding/maintenance cost (HC) and the transaction cost(TC_3) as represented in (4)

$$U_1(t) = V(t) - HC(t) - TC_1 + S(t) \tag{1}$$

$$U_2(t) = Max\{U_g, U_r\}_t \tag{2}$$

$$U_g = HC(t) - V(t) - TC_2 + R \tag{3}$$

$$U_r = Re(V(t)) + HC(t) - V(t) - TC_3 \tag{4}$$

Equations (1) to (4) depict the economic rationality behind sharing/gift/resale decision in general. We can further define the product valuation, the holding cost, and the sharing income according to time as follows:

$$V(t) = A\left(\frac{e^k}{1+i}\right)^t \tag{5}$$

$$HC(t) = \frac{C}{\ln(1+i)}\left[1 - (1+i)^{-t}\right] \tag{6}$$

$$S(t) = V * E\left[\frac{S}{V}\middle|x = t\right] = V(t)e^{\frac{1}{2}\sigma t^2} \tag{7}$$

$$Re(V) = V(t) \tag{8}$$

In Eq. (5), k is a kind of value power which shows the change of value. Value decreases with time when $k < 0$ while value increases when $k > 0$. In (6), we assume that unit holding cost is a constant. In Eq. (7), we assume that at any time, $S(t)/V(t)$} is a random variable which has a logarithmic normal distribution with parameters 0 and σt. It's natural to consider σt increases when t increases because of the characteristic of volatility. So we could let $\sigma t = st$ with $s > 0$. In Eq. (8), we assume $Re(V)$ is a normal random variable with parameters $V(t)$ and $\sigma't$. Equations (5) to (8) are still very general, and can be fine-tuned according to commodity types in different industries or applications. We use these equations to facilitate the calculations and to demonstrate the boundaries and conditions in different economic forms.

Theorem 1. At any time point if $\alpha = Re(V(t)) - TC_3 + TC_2 - R > 0$, resale is more preferable than gift giving. Otherwise, the good owner would rather like to make a donation.

Proof. It can be easily proved by comparing U_r with U_g where α stands for the difference between these two possible owner's utilities.

From now on we use U_2 to represent the utility of detaching the good, where $U_2(t) = Max\{U_g, U_r\}_t$. If we consider U_1 and U_2 as two substitute choices for the owner, which means if U_1 surpasses U_2, the owner intends to share. Otherwise, if U_2 is greater than U_1, he/she prefers to detaching the good either as a gift or as a resale item. His/her maximization problem for perfect complements can be modeled as follows:

$$MaxU\{x, y\}_t \tag{9}$$

subject to:

$$x_t U_1 + y_t U_2 = U\{x, y\}_t \tag{10}$$

$$x_t + y_t = 1 \tag{11}$$

$$x_t \in \{0, 1\} \tag{12}$$

$$y_t \in \{0, 1\} \tag{13}$$

From Eqs. (5) to (7), We can form the following intermediate formulas regarding the remaining value of the good, the inventory cost/credit, and the sharing income.

$$V'(t) = A \left(\frac{e^k}{1+i} \right)^t *[k - \ln(1+i)] \tag{14}$$

$$V''(t) = A \left(\frac{e^k}{1+i} \right)^t *[k - \ln(1+i)]^2 \tag{15}$$

$$HC'(t) = C(1+i)^{-t} \tag{16}$$

$$HC''(t) = -C\ln(1+i)(1+i)^{-t} \tag{17}$$

$$S'(t) = S(t) \left[\sigma_t \sigma'_t + k - \ln(1+i) \right] \tag{18}$$

and draw the graphs of U_1 and U_2 respectively in the same coordinate system in order to determine whether $x = 1$ or $y = 1$. Alternatively, we can also observe the function $U_1 - U_2$ directly to find the even point.

In detail:

$$U_1(t)' = V' - HC' + S' \tag{19}$$

$$U_1(t)'' = V'' - HC'' + S'' \tag{20}$$

$$U_1(0) = A - TC_1 \tag{21}$$

$$U_2(t)' = HC' - V' \tag{22}$$

$$U_2(t)'' = HC'' - V'' \tag{23}$$

$$U_2(0) = -A + R - TC_2 \tag{24}$$

Theorem 2. If $k > \ln(1 + i)$, the owner would be more likely to prefer to sharing/holding the good than giving it out as a gift.

Proof. Because the value of the good always increases, the owner would consider keeping the ownership of the good as appreciation instead of depreciation.

So we will discuss under the circumstance $k < \ln(1 + i)$, which makes $V' - HC' = A[k - \ln(1+i)]\left(\frac{e^k}{1+i}\right)^t - C(1+i)^{-t}$ lower than zero in that case.

In order to observe U_1 and U_2 more conveniently and clearly, the function $[U_1 - U_2](t)$ would be discussed in the following. Let

$$[U_1 - U_2](t) = \phi(t) - \theta \tag{25}$$

where

$$\phi(t) = S(t) + 2(V - HC) \tag{26}$$

$$\theta = TC_1 + R - TC_2 \tag{27}$$

Then the condition $U_1 > U_2$ is equivalent to $\phi(t) > \theta$, which is also equivalent to $S(t) + 2(V - HC) > TC_1 + R - TC_2$.

From the analysis above, we know that $V - HC$ is a monotonous decreasing convex function based on the fact that $V' - HC' < 0$ and $V'' - HC'' > 0$. On the other hand, S' is linear monotonous increasing function of t and when $t = \frac{\ln(1+i)-k}{s^2}$, $S' = 0$. So $\phi' = S' + 2 V' - 2HC'$ is a monotonous increasing function from negative to positive with the unique zero point t_0. In other words, when $0 < t < t_0$, $\phi(t)$ decreases with t, while when $t > t_0$, $\phi(t)$ increases with t, which shows that t_0 is the minimum point of the function $\phi(t)$.

Theorem 3. If the minimum of the function $\phi(t)$ is larger than or equal to θ which is the same with $\phi(t_0) \geq \theta$, then $U_2(t)$ won't exceed $U_1(t)$, which means the owner will share the good until the end of product life cycle.

Proof. From the perspective of $[U_1 - U_2](t)$, we can get the minimum of the function $[U_1 - U_2](t)$ is not below zero given those conditions, which result in $U_1 - U_2 \geq 0$ is correct for all t. In other words, $U_1(t) \geq U_2(t)$ is always correct.

Let t^* be the first point which satisfies that $\phi(t^*) = \theta$, then:

Theorem 4. If the minimum of the function $\phi(t)$ is smaller than θ, which is the same with $\phi(t_0) < \theta$, then t^* is the potential separation point, which means the owner will give the good away at time t^* indeed if the gray area is larger than the second red area, otherwise the owner will share it until T. However, if $t_0 \geq T$, which indicates there is no red area, then t^* is surely a separation point.

Proof. Because of the former analysis of the property of the function $\phi(t)$, we know there are two intersections for us to consider when θ is within the range of function value. In detail, t^* is sure to be achieved and it's more important than the second one because t^* is potential giving point while the second one is not. We just need it to determine whether to give the good at the time t^* by comparing the red area after the second intersection and the gray area before it. Therefore, its existence and value of second intersection are not significant. If the second separation point doesn't exist when $t_0 \geq T$, then we can consider red area as zero, which is necessarily less than gray area (Fig. 3).

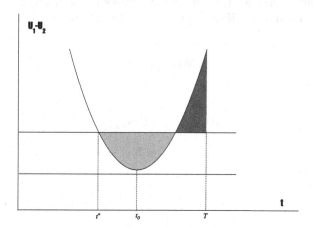

Fig. 2. Temporal ownership boundary when $t_0 < T$

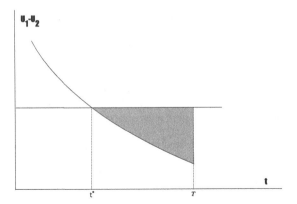

Fig. 3. Temporal ownership boundary when $t_0 > T$

3.3 An Alternative Model

Because it's possible for the owner to share the good for a certain time period and give it as a gift in the future, we may consider x and y as the tendency of sharing and gift-giving respectively. In this case, we shall abandon constraints (10), (11), (12) and make x and y continuous. In other words, we will take the two different statuses of a certain good, which are sharing and giving, as two competitors in order to find the result of the battle between sharing and giving.

If we consider the total remaining life of the good as T and when the owner gives it as a gift in a future time point t, x represents the proportion of sharing as x = t/T, and y represents the proportion of gift-giving y = 1 − x. The original problem from (3) to (7) becomes:

$$MaxU\{x, y\} \tag{28}$$

subject to:

$$xU_1 + yU_2 = U\{x, y\}_t \tag{29}$$

$$x + y = 1 \tag{30}$$

$$x = \frac{t}{T}, x \in [0, 1] \tag{31}$$

$$y = 1 - \frac{t}{T}, y \in [0, 1] \tag{32}$$

$$t \leq T \tag{33}$$

$$U_1(t) = V(t) - HC(t) - TC_1 + S(t) \tag{34}$$

$$U_2(t) = HC(t) - V(t) - TC_2 + R \tag{35}$$

$$V(t) = A\left(\frac{e^k}{1+i}\right)^t \tag{36}$$

$$HC(t) = \frac{C}{\ln(1+i)}\left[1 - (1+i)^{-t}\right] \tag{37}$$

(36) states the intrinsic value of the good at a future time t from time point 0. In order to calculate V, we take into consideration depreciation that consists of the initial value of the asset and its estimated "life". (37) represents the accumulated value of released inventory cost as an "inventory holding credit".

From the perspective of (30), MRS = −1, while from the perspective of the linear utility function (29), $\frac{MU_x}{MU_y} = -\frac{U_2}{U_1}$ as per microeconomics principles, MRS = MU_x/MU_y indicates $U_1(t) = U_2(t)$ is the condition for the optimal solution, which is the same condition that we discussed earlier.

What's more, for the purpose of maximizing the utility, we need to find the point t^* which makes $\frac{\partial U}{\partial t}(t^*) = 0$. We can simplify that $\frac{dU}{dt} = \frac{1}{T}[U_1 - U_2] + \frac{1}{T}[U_1' - U_2'] + U_2'$. And it's easy to prove the solution of that equation is exactly t^*.

4 Economic and Managerial Implications

In general, as can be observed from Fig. 2, P_s and P_g are separable by t^*, which could be affected by the gift rewards R and sharing income S. The reward usually has something to do with tax deduction, while sharing income usually directly relates to cost and sharing platform.

The impact from the sharing income is that if only s increases, then t_0 decreases and t^* increases finally, even making giving not to happen as shown in Table 1.

Table 1. Analysis of temporal ownership boundary according to various s

PAR	Th2	Th3	Th4	Th5	Th6	Th7	Th8	Th9
A	10	10	10	10	10	10	10	10
i	0.1	0.1	0.1	0.1	0.1	0.1	0.1	0.1
k	0.1	0.09	0.09	0.09	0.09	0.09	0.09	0.09
C	2	2	2	2	2	2	2	2
s	0.01	0.02	0.03	0.04	0.05	0.06	0.07	0.08
R	10	10	10	10	10	10	10	10
TC_1	1	1	1	1	1	1	1	1
TC_2	2	2	2	2	2	2	2	2
T	20	20	20	20	20	20	20	20
θ	9	9	9	9	9	9	9	9
t^*	NA	6.8	6.9	6.9	7	7.2	7.4	7.7

Table 2. Analysis of temporal ownership boundary according to various θ

PAR	Th2	Th3	Th4	Th5	Th6	Th7	Th8	Th9
A	10	10	10	10	10	10	10	10
i	0.1	0.1	0.1	0.1	0.1	0.1	0.1	0.1
k	0.09	0.09	0.09	0.09	0.09	0.09	0.09	0.09
C	2	2	2	2	2	2	2	2
s	0.05	0.05	0.05	0.05	0.05	0.05	0.05	0.05
R	7	8	9	10	10	10	11	11
TC_1	1	1	1	1	1	2	2	3
TC_2	2	2	2	2	1	1	1	1
T	20	20	20	20	20	20	20	20
θ	6	7	8	9	10	11	12	13
t^*	8.7	8.1	7.5	7	6.6	6.1	5.7	5.2

The impact from gift reward can be found by examining R. We can write R as $\lambda V(t) + R_0$. As a result, there is an adjustment from 2 to $2 - \lambda$ in the coefficient of the V(t) and an added constant in the function of $U_1 - U_2$. However, this adjustment only creates small changes. If only R increase, then t^* decreases finally. And vice versa. The change in TC_1 and TC_2 also influences the value of t^* via changing the minimum of the function $[U_1 - U_2](t)$. The integrated effect of $R + TC_1 - TC_2$ is shown in Table 2.

From the above analysis, we can conclude that:

Theorem 5. The decrease in s has the same effect with the increase in $R + TC_1 - TC_2$, which would both cause the decrease in t^*. Furthermore, inverse change in $R + TC_1 - TC_2$ and S' would strengthen their effect while synchronized change in $R + TC_1 - TC_2$ and S' would counteract their respective effects.

From the perspective of the owner of the good, the higher σ_t would bring higher income but the high income would discourage the owner to give the good as a gift according to our results. However, the high σ_t would also make the good less competitive at the same time. As a result of that, there is an equilibrium in the price setting for the sharer to obtain the maximum income and meanwhile it wouldn't eliminate the possibility for gift giving.

From the perspective of government, to increase the reward by enhancing the tax deduction is a good method to encourage people to give the good away as soon as possible. But it's difficult to promote tax deduction without limiting cap, which is the most common current practice. As it can be seen from our results, the game between the sharing income and gift economy reward would play a key role in the owner's decision. Adjusting the tax deduction corresponding to the price of sharing market would make the reward more effective.

From the perspective of firms that promote gift economy, normally these firms are considered non-profit. There exist many challenges that they have to overcome, such as the lack of supply, the increasing demand of charity, the lack of understanding of both donors and receivers, and sometimes the financial difficulties to operate the platform and to reduce the transaction cost. Our result shows that by reducing the transaction cost, the time for people to give the good as gift would be brought forward, which means the platform would receive the goods earlier. Even for those who will not donate the goods, the reduced transaction cost and increased rewards would give them the motivation to donate. Our results also show that today's ever booming sharing and resale economy, because of the reduced transaction cost to share and to resell along with increasing sharing and resale income thanks to the Internet, actually shrink the already small market size for the gift economy. We prove that in order to boost the spirit of good-will in our society, the government and the charity organizations must come up with new models or effective taxation incentives to struggle with the increasing income of sharing or reselling in order to encourage the gift giving, like revising the tax deduction corresponding to the price of sharing market.

From the perspective of the sharing or reselling platform, reducing the transaction cost by new technology would help them receive more goods from the owner. And the goodwill for their efficiency with distribution of the goods would attract more people to share or give their goods because they may believe that platform could help them fulfill their purpose. In that way, owing to the double-sided model, platform would benefit from the increasing sharer/donors by attracting more receivers, which would in turn enhance the volatility of the platform. Nevertheless, transaction cost is the profit of platform, which means the transaction cost would not decease without limit. Compared with that in gift economy, our results show that the difference between two kinds of transaction cost could be utilized by government or charity organizations to encourage people to give their good as a gift.

5 Conclusion

We investigate the temporal ownership boundary that exists in the sharing economy. We study the temporal factors including the inventory holding cost, the potential collaborative income, and individual utility from consumption with various time stamps. We define the temporal ownership boundary as the limit when the owner is indifferent to transferring the ownership from its current in-usage or sharing status. We base our analysis on two variations of substitute modeling and consider the properties of social welfare by incorporating the utility functions of different players. We find that there exist various conditions when this boundary may lean towards sharing, gift giving or reselling. We show that both individual utility and total social welfare can be optimized by adjusting the incentives, the transaction costs, and eventually the time of ownership transfer of the goods. Our results bring meaningful and interesting insights to today's sharing, gift, and resale platform companies on how to improve the efficiency and competitiveness.

Thanks to the rapid development of various online social networks and recommender systems, today's consumers are able to gain access to information instantly, to communicate with other consumers conveniently, and to enjoy low cost online c2c transactions. The Internet has enabled the booming of the three emerging economic forms that we have discussed in this research. For future research, we foresee many variations and new economic models based on the temporal ownership boundary. For example, in reality the parameter of $k \leq \ln(1 + i)$ in our model happens to appear more frequently than the ones when k > ln(1 + i). However, the existing reward function loses its influence in the case of $k \le \ln(1+i)$, which means an alternative reward mechanism should be designed.

References

1. Ausubel, L.M., Cramton, P.: Vickrey auctions with reserve pricing. Assets, Beliefs, and Equilibria in Economic Dynamics. Springer Berlin Heidelberg, pp. 355–367 (2004)
2. Belk, R.: You are what you can access: sharing and collaborative consumption online. J. Bus. Res. **67**(8), 1595–1600 (2014)
3. Botsman, R., Rogers, R.: What's Mine is Yours: How Collaborative Consumption is Changing the Way We Live. Collins, London (2011)
4. Cheal, D.: The Gift Economy. Routledge, Abingdon (2015)
5. Cohen, B., Kietzmann, J.: Ride on! Mobility business models for the sharing economy. Organ. Environ. **27**(3), 279–296 (2014)
6. Garratt, R., Trotger, T.: Speculation in standard auctions with resale. Econometrica **74**(3), 753–769 (2006)
7. Hamari, J., Sjöklint, M., Ukkonen, A.: The sharing economy: Why people participate in collaborative consumption. J. Assoc. Inf. Sci. Technol. **67**(9), 2047–2059 (2016)
8. Malhotra, A., Van Alstyne, M.: The dark side of the sharing economy and how to lighten it. Commun. ACM **57**(11), 24–27 (2014)
9. Marcoux, J.S.: Escaping the gift economy. J. Consum. Res. **36**(4), 671–685 (2009)
10. Mas, A.: Labour unrest and the quality of production: evidence from the construction equipment resale market. Rev. Econ. Stud. **75**(1), 229–258 (2008)
11. Schor, J.: Debating the sharing economy. Great transition initiative (2014)
12. Sundararajan, A.: From Zipcar to the sharing economy. Harvard Bus. Rev. **1** (2013)
13. Zervas, G., Proserpio, D., Byers, J.: The rise of the sharing economy: estimating the impact of Airbnb on the hotel industry. Boston University School of Management Research Paper (2013–2016) (2015)

Optimal Pricing and Workforce Composition for Service Delivery Using a Hybrid Workforce (Research in Progress)

Su Dong[1]([⊠]), Monica S. Johar[2], and Ram L. Kumar[2]

[1] School of Business and Economics, Fayetteville State University,
Fayetteville, US
sdong@uncfsu.edu
[2] Belk College of Business, UNC Charlotte, Charlotte, US
{msjohar,rlkumar}@uncc.edu

Abstract. Innovative work arrangements are increasingly being enabled by technology. On-demand technology-enabled marketplaces for a variety of skills are becoming popular. Examples include topcoder.com and odesk.com. This paper focuses on how organizations can use such on-demand marketplaces along with in-house workers to staff dynamic knowledge-intensive service environments. An economic analysis of prices paid to on-demand workers is presented. Such an analysis can be further developed to characterize the optimal workforce composition of the in-house workforce.

1 Introduction

The dynamic technology-driven business environment of today is facilitating innovative work arrangements for service delivery using a distributed workforce. Examples of such services include security (www.qualys.com) and business services such as customer service (www.salesforce.com). High quality customer service is extremely important, and in some settings this requires a skilled (knowledge-intensive) workforce. Instead of relying on in-house (IH) workers to provide high quality service, technology is facilitating innovative work arrangements using an on-demand (OD) workforce (The Economist 2015). Some people argue that hybrid workforces are likely to be "the new normal" (Kasriel 2015). Hence, it is vital for organizations today and in the future to explore work arrangements that employ a hybrid workforce consisting of full-time and OD workers. This research focuses on pricing and workforce composition in such hybrid workforce settings.

We examine pricing and composition of hybrid workforces in the context of knowledge intensive service environments that are characterized by a workforce with competencies in multiple skills, and significant heterogeneity in worker competence (Dong et al. 2012). We focus on knowledge intensive service delivery environments since OD workforces are increasingly available for knowledge intensive tasks (The Economist 2015). Service tasks could range from several minutes to a several hours. Assignment of tasks to skilled workers is a key determinant of service performance.

M. Fan et al. (Eds.): WeB 2016, LNBIP 296, pp. 67–73, 2017.
https://doi.org/10.1007/978-3-319-69644-7_6

An assignment strategy that assigns tasks to the first available worker, could reduce wait times, but result in high service task completion times.

Prior research (Aksin et al. 2007) as well as practitioner articles (Rugaber 2013) have recognized the potentially important role that part-time workers play in service provision. However there is limited research on service using hybrid workforces., Related research in service operations, and service science (Aksin et al. 2007; Dong et al. 2011) has focused on service provision by a full-time workforce. There is some research focusing on non-routine service tasks (Cil et al. 2013) involving part-time workers (Rugaber 2013). In today's business environment, several marketplaces for part-time service workers exist. Examples include www.topcoder.com, and www.odesk.com, which provide access to an on-demand workforce for software development and legal services. While there is some research on such service marketplaces (Cil et al. 2013; Bcaon et al. 2010), there is no research focusing on pricing and composition of hybrid workforces. Our focus, in this paper, is on software development service tasks, though our research can apply to other types of tasks as well.

The service environment studied in this paper combines an IH workforce with independent OD workers who bid for tasks. Service tasks are associated with service levels. Timely completion of tasks is important in order to minimize penalties and maximize net revenue. We focus on the following research question: what is the best (optimal) pricing and workforce composition for a hybrid workforce? An economic analysis of conditions under which an organization uses the on-demand market place and optimal price paid to OD workers is presented. The composition (range of worker service times) of the OD workforce that the organization can attract with this price is also discussed.

2 Service Delivery Using a Hybrid Workforce

This section develops an analytical model of a service environment that uses a hybrid workforce. This model aims to improve our understanding of interrelated pricing, workforce and service level agreements impact the performance of the firm.

Service tasks are assigned to workers in order to generate business value. Tasks can be assigned to either OD workers or IH workers. IH and OD workers vary in terms of their availability, compensation, and skill levels. These knowledge workers may not handle service tasks immediately upon arrival, though delay in task completion may result in penalties outlined in SLAs.

IH workers are paid a fixed salary. OD workers are screened by the firm based on their skill levels, reputation (based on past performance) and experience. Tasks along with their pre-determined payment price and service delivery requirements are posted to a market of OD workers (e.g., topcoder.com, odesk.com). OD workers place (yes/no) bids for these tasks based their utility functions. Each worker's utility function typically depends on the payment for the task, the learning value from the task, and the opportunity cost related to completing the task. Value of learning and the opportunity cost, in turn, depend on worker skills and market demand for those skills. Each OD worker who bids has no knowledge of other bids or other tasks arriving in future time periods. When there are multiple bids for a task, the firm makes an assignment that maximizes its profit objective. A worker can bid on multiple tasks in any given time period, but is assigned to only one task at a time.

3   Model Formulation

The firm's objective is to maximize business value by completing service tasks over a planning horizon (T). In line with prior research on managing IT service tasks, we assume that the arrival rate of tasks follows a Poisson distribution (with a mean λ), and the tasks arriving in each time period are independent of each other. Each task has an associated revenue of $\bar{R}$. Tasks can be completed by IH or OD workers. At the beginning of the planning horizon the firm hires a heterogeneous IH workforce of size K (decision variable) with the average service time $1/\mu_{IH}$ (decision variable). We assume that the actual service time for the IH workforce follows a general distribution. Similarly, the average service time for OD workers is given by $1/\mu_{OD}$. IH workers are paid a fixed compensation $h(1/\mu_{IH})$ for the entire planning horizon based on their competence (service time). On-demand workers accept tasks based on their individual utility function (U_{OD}) and the payment price P_{OD} (decision variable) set by the firm.

4   On-Demand Marketplace Characterization

The net utility derived by an on-demand worker is function of the direct compensation (P_{OD}) associated with task completion, opportunity cost ($g(\tau, 1/\mu_{OD})$) which is a function of the average service time of the OD worker ($1/\mu_{OD}$) and external market demand (τ). In addition, we also assume that the OD worker may derive some utility form learning associated with task completion ($f(\theta, \psi, 1/\mu_{OD})$). The value of learning is function of market price for skills (ψ), the task related learning potential (θ) and the service time ($1/\mu_{OD}$) of the worker. Hence, we model the utility function of the OD worker as:

$$U_{OD} = P_{OD} + f(\theta, \psi, 1/\mu_{OD}) - g(\tau_{kt}, 1/\mu_{OD}) \qquad (1)$$

Lemma 1. OD workers with a service time greater that $1/\mu_{OD}^m$ will chose to participate in the OD marketplace.

The value of learning $f(\theta, \psi, 1/\mu_{OD})$ increases with market value of skills and the task related learning potential. Since, higher service times are indicative of less competent workers we assume that the opportunity cost decreases with the average service time ($1/\mu_{OD}$) and increases with external market demand. From Eq. (1), the price for the marginal worker is given by,

$$P_{OD}^m = g(\tau_{kt}, 1/\mu_{OD}^m) - f(\theta, \psi, 1/\mu_{OD}^m). \qquad (2)$$

Since the payment to the OD worker strictly decreases with increase in service time, it follows that $\frac{dP_{OD}}{d(1/\mu_{OD})} < 0$.

Lemma 1 is illustrated in Fig. 1 where the individual rationality constraint for the OD worker is binding at the marginal curve. Workers with a service time greater than $1/\mu_{OD}^m$ will chose to participate in the OD market place.

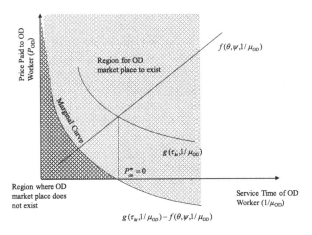

Fig. 1. Illustrative example of operating region for the OD marginal worker.

From the firm's perspective the service delivery environment can be approximated to a M/G/K queue with an arrival rate of λ and general service time distribution for K (decision variable). Hence, in the absence of OD market place, the average waiting time for a task in the queue is given by,

$$E[W^{M/G/K}] = 0.5\,(CV^2 + 1)E[W^{M/M/K}] \tag{3}$$

Here CV^2 is the coefficient of variation of the service time distribution for IH workers. In M/M/K the server utilization $\rho = \lambda/(K\mu_{IH})$. The probability that an arriving customer is forced to join the queue (i.e., all servers are occupied is given by):

$$C(K, \lambda/\mu_{IH}) = \frac{1}{1 + (1 - \rho)(K!/(K\rho)^K)(\sum\limits_{m=0}^{K-1} (K\rho)^m/m!)} \tag{4}$$

The waiting time in the queue for M/M/K is given by,

$$E[W^{M/M/K}] = \frac{C(K, \lambda/\mu_{IH})}{K\mu_{IH} - \lambda} \tag{5}$$

Combining Eqs. 3, 4 and 5 we get the waiting time in the queue for the M/G/K queue.

The total task completion time is the sum of the waiting time and the service time. The maximum service time for each task before the firm incurs penalty of δ dollars per unit time is defined as Individual Based SLA (IBSLA). The IH workforce is hired at the beginning of the planning horizon for the entire planning horizon, hence the additional cost of using an IH worker is,

$$\max(E[W^{M/G/K}] + 1/\mu_{IH} - IBSLA, 0)\,\delta. \qquad (6)$$

Tasks can also be posted in the OD marketplace for bidding along with their pre-determined payment price (P_{OD}) and service delivery requirements. OD workers with a positive net utility are assumed to bid for tasks at the posted payment price. Each OD worker who bids has no knowledge of other bids or other tasks arriving in future time periods.

Since, an OD worker's service time can also exceed IBSLA, the additional cost of using an OD worker for a task is given by,

$$\max(E[W^{M/G/K}] + 1/\mu_{IH} - IBSLA, 0)\,\delta. \qquad (7)$$

Proposition 1. characterizes the maximum price offered by the firm for tasks posted in the OD market place.

Proposition 1a. If the total average service time of an OD workers ($1/\mu_{OD}$) is less than the IBSLA and the average total service time for the IH worker ($E[W^{M/G/K}] + 1/\mu_{IH}$) is greater than the IBSLA, the firm posts the tasks in the OD market place for a maximum price ($P_{OD}^{\max}$) equal to $(E[W^{M/G/K}] + 1/\mu_{IH} - IBSLA)\,\delta$.

In this service environment, the firm does not incur any additional cost savings as long as the total service time is less than IBSLA. Hence, under the conditions described in Proposition 1a it follows that maximum market price will be equal to the additional cost of using the IH workers (Eq. 6). This is illustrated in Fig. 2.

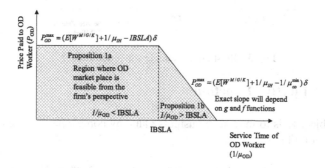

Fig. 2. Illustrative example for Proposition 1.

Proposition 1b. If the average service time of an OD workers ($1/\mu_{OD}$) is greater than the IBSLA but less than the average total service time for the IH worker ($E[W^{M/G/K}] + 1/\mu_{IH}$), the firm posts the tasks in the OD marketplace for a maximum price ($P_{OD}^{\max}$) equal to $P_{OD}^{\max} = (E[W^{M/G/K}] + 1/\mu_{IH} - 1/\mu_{od})\,\delta$.

In this case (Proposition 1b) the maximum payment price is proportional to the potential time savings due to the use of an OD worker instead of an IH worker. However, maximum payment price decreases with increase in $1/\mu_{OD}$ because the service time of the OD worker also exceeds IBSLA. Illustrated in Fig. 2.

Proposition 1c. The firm does not find it optimal to post the tasks to the OD marketplace if the average total service time for the IH worker is less than the IBSLA or the time taken by the OD worker is greater than the total service time taken by the IH worker.

The above result follows from the fact that the OD market place is only useful under conditions when the IH workforce exceeds IBSLA and service times higher than those available in the OD market place.

Proposition 2. The service times of participating OD workers is in the range $\mu_{OD} \in [\mu_{min}, \mu_{max}]$ where, μ_{min} and μ'_{OD} are solutions to $(E[W^{M/G/K}] + 1/\mu_{IH} - IBSLA)\delta = g(\tau_{kt}, 1/\mu_{OD}^{min}) - f(\theta, \psi, 1/\mu_{OD}^{min})$ and $(E[W^{M/G/K}] + 1/\mu_{IH} - 1/\mu'_{OD})\delta = g(\tau_{kt}, 1/\mu'_{OD}) - f(\theta, \psi, 1/\mu'_{OD})$, respectively and $1/\mu_{OD}^{max} = Max(1/\mu'_{OD}, 0)$.

The intuition for the above result is as follows. If the service time of the OD worker is smaller than a critical value $1/\mu_{min}$, then the marginal price of an OD worker makes the firm less profitable. Conversely, if the time take by the OD worker is greater than the critical $1/\mu_{max}$, the firm finds it optimal to use the IH worker since there is no longer any cost saving from using the OD workforce. This is illustrated in Fig. 3.

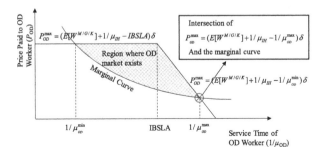

Fig. 3. Illustrative example for Proposition 2.

5   Firm's Objective Function

The firm's objective is to maximize business value by completing service tasks over a planning horizon (T).

$$J = (\bar{R} - \max((E[W^{M/G/K}] + 1/\mu_{IH} - IBSLA), 0))\delta\min(\lambda T, KT\mu_{IH}) - h(P_{IH}, 1/\mu_{IH})K$$
$$+ (\bar{R} - P_{OD} - \max((1/\mu_{OD} - IBSLA, 0)\delta)(\lambda T - \min(\lambda T, KT\mu_{IH}))$$

$$(12)$$

Here, $\bar{R}$ is the price per unit task and $h(1/\mu_{IH})$ is the price function for the IH workforce. From Lemma 1 and Propositions 1 and 2, we can see that optimal price

(P_{OD}^* curve) for the OD market place will follow the marginal curve with the firm extracting all the consumer surplus for $1/\mu_{OD} \in [1/\mu_{min}, 1/\mu_{max}]$.

Hence, the objective can be re-written as,

$$\underset{K,\mu_{IH},\mu_{OD}}{Max} \; J = \underset{K,\mu_{IH},\mu_{OD}}{Max} \; \{(\bar{R} - \max(E[W^{M/G/K}] + 1/\mu_{IH} - IBSLA))\delta\min(\lambda T, KT\mu_{IH})$$
$$- h(P_{IH}, 1/\mu_{IH})K + (\bar{R} - (g(\tau_{kt}, 1/\mu_{OD}) - f(\theta, \psi, 1/\mu_{OD}))$$
$$- \max(1/\mu_{OD} - IBSLA, 0)\delta(\lambda T - \min(\lambda T, KT\mu_{IH}))\} \qquad (13)$$

For further analysis we need to assume functional forms for f, g, and h and find the solution for K* and $1/\mu_{IH}^*$.

6 Conclusion

The emergence of on-demand service marketplaces is a relatively new phenomenon. As the range of services available on such marketplaces increases, organizations could explore innovative uses of hybrid workforces. This paper addresses this interesting new work paradigm by presenting a model of service delivery that leverages in-house workers and on-demand marketplaces. This research in-progress paper has focused on optimal pricing. Our results illustrate how organizations arrive at optimal prices. Optimal workforce size and composition (average service time) can be arrived at analytically. We will present these results at the conference.

References

Aksin, Z., Armony, M., Mehrotra, V.: The modern call centre: a multidisciplinary perspective on operations management research. Prod. Oper. Manag. **16**(6), 665–688 (2007)

Bacon, D.F., Bokelberg, E., Chen, Y., Kash, I.A., Parkes, D.C., Rao, M., Sridharan, M.: Software economies. In: Proceedings of the FSE/SDP Workshop on Future of Software Engineering Research, pp. 7–12 (2010)

Cil, E.B., Allon, G., Bassamboo, A.: Large-scale service marketplaces: the role of the moderating firm. www.kellog.northwestern.edu. http://www.kellogg.northwestern.edu/faculty/allon/htm/research/service_matketplace.pdf. Accessed 29 July 2013

Dong, S., Johar, M.S., Kumar, R.L.: A benchmarking model for management of knowledge-intensive service delivery networks. J. Manag. Inf. Syst. **28**(3), 127–160 (2011)

Kasriel, S.: The hybrid workforce: why combining onsite, remote, and freelance resources is becoming the new normal. http://venturebeat.com/2015/06/02/the-hybrid-workforce-why-combining-onsite-remote-and-freelance-resources-is-becoming-the-new-normal/. Accessed Aug 2015

Rugaber, C.S.: Temporary Jobs Becoming a Permanent Fixture. www.usatoday.com. http://www.usatoday.com/story/money/business/2013/07/07/temporary-jobs-becoming-permanent-fixture/2496585/. Accessed 29 July 2013

The Economist: The on-demand economy workers on tap. http://www.economist.com/news/leaders/21637393-rise-demand-economy-poses-difficult-questions-workers-companies-and. Accessed Aug 2015

The Performance Evaluation of Machine Learning Classifiers on Financial Microblogging Platforms

Tianyou Hu[✉] and Arvind Tripathi

University of Auckland, Auckland, New Zealand
{t.hu,a.tripathi}@auckland.ac.nz

Abstract. As technological advancements facilitate democratization of knowledge, Microblogging platforms are vying to become the premier source of knowledge and are competing with news outlets. A huge number of messages is generated on different microblogging platforms. In financial markets, microblogging websites, such as StockTwits, have become a rich source for amateur investors, which make them ideal sources for market sentiment analysis. Indeed, StockTwit has been widely used by researchers for sentiment analytics and market predictions. However, the quality of the sentiment analysis is highly dependent on the machine learning classifiers used as well as the preprocessing of data. In this study, we compare the performance efficiency of different machine learning classifiers on the user-generated content on StockTwits. We find that Logistic Regression Classifier performs best in a 2-way classification of Stock-Twits data. Our results report better classification accuracy than a similar research using data from Twitter. We have discussed managerial implications of our results.

Keywords: Social media · Microblogging machine · Learning sentiment analysis · User-generated content · Stock market

1 Introduction

Social media, especially microblogging services are becoming popular sources of information in almost all domains. For example, millions of Tweets are generated on Twitter everyday. Users create, share and discuss information on various topics, from personal life, and healthcare problems to societal issues and politics. Financial analysis and investment strategies, which used to be the limited to domain experts, is now provided by retail investors on social media [1]. The quality of information available on social media platforms is comparable to expert opinions. In fact, many studies have established connections between sentiments on social media platforms and market returns [1–3]. Many studies have analyzed tweets from Twitter but since Twitter covers a very broad range of topics, it's difficult to filter and choose the right Tweets concentrating on the desired topic. We argue that domain specific microblogging platforms, such as StockTwits for stock market provide a better data source to study discussions and analyze market sentiments.

© Springer International Publishing AG 2017
M. Fan et al. (Eds.): WeB 2016, LNBIP 296, pp. 74–83, 2017.
https://doi.org/10.1007/978-3-319-69644-7_7

In recent years, researchers have shown the effect of sentiments derived from microblogging platforms on stock markets [3, 4]. Social media users use microblogging services to share their opinion about stock markets. This huge amount of data on microblogging platforms like StockTwits, is a treasure trove for market analysts and becomes a new market sentiment indicator and competes with the one based on traditional sources (newspaper, online news media or blogs written by experts). Furthermore, the short length of each message (maximum 140 characters per message) and the use of cashtags (an identifier like hashtag but starts with '$') make it a less noisy and easier to analyze. Furthermore, high frequency of content creation by users also allows analysts to track user behavior at a different level, in real-time, during trading.

Given the untrusted content, it's very challenging for an average person to process the huge amount of data and estimate market sentiments. These shortcomings can be addressed by using machine learning techniques. There has been increasing interest in stock market predictions using various machine learning techniques. Different machine learning algorithms have been used to classify messages into different sentiment groups. However, we are yet to understand classification efficiency of these algorithms for analysing messages from a microblogging platform. In this research, we compare the classification performance of different classifiers used for classifying posts on a microblogging platform StockTwits.

Section 2 reviews the related literature on feature selection and sentiment analysis methods. Section 3 describes the data used in this research. Section 4 explains the machine learning classifiers used in this research. Section 5 presents the results. We conclude with a discussion in Sect. 6.

2 Literature Review

In literature, many approaches have been used to conduct sentiment analysis in social media.

Researchers have used various pre-defined dictionaries and machine learning classifiers to extract user sentiments from social media messages and articles in different context. To deal with this issue, Loughran and McDonald [5] compared the four most widely used dictionaries, which are Henry [6], Harvard's General Inquirer (GI), DICTION, and L&M [7]. Each dictionary has its expertise, but the L&M is better than the rest three dictionaries in financial context for the following two reasons. First, the L&M dictionary does not miss common positive and negative words, which makes it more comprehensive than the rest. Second, the L&M dictionary was created for financial context analysis. It has been shown that L&M does really poor in short message classification in comparison with machine learning classifiers [8]. Thus, we will only compare machine-learning classifiers in this study.

Regarding the state of the art for machine learning classification in financial markets, Antweiler and Frank [9] came up with a novel idea to compute bullishness index using computational linguistics method and showed that stock messages can predict market volatility. Bollen et al. [4] measured collective mood state in term of two states (positive vs negative) and 6 dimension (Cal, Alert, Sure, Vital, Kind and Happy) from Twitter data using OpinionFinder and Google Profile of Mood States, and found an accuracy of 86.7% in predicting the directional changes in the closing price of Dow Jones Industrial Average. Sprenger et al. [10] collected Twitter messages containing cashtags of S&P 100 companies and classified each message using Naïve Bayes (NB) trained with a set of 2,500 tweets. Results demonstrated that bullishness index is correlated with the abnormal return and message volume is associated with trading volume. Oh and Sheng [2] collected data from StockTwits for three months. The messages were classified by a bag of words approach which applied a machine learning algorithm J48 classifiers. They argued that the sentiments appear to have strong forecasting power over the future market directions. Tirunillai and Tellis [11] collected data from consumer reviews and classified the reviews using NB and Support Vector Machine (SVM). Results showed that negative UGC has a significant negative effect on abnormal returns with a short "wear-in" and long "wear-out" effects, positive UGC has no significant effect on these metrics. Oliveira et al. [12] collected data from Stock-Twits for 605 trading days. Messages were counted as "bullish" if they contain the words "bullish", same logic was applied to messages containing "bearish" words. In contrast with previous studies, they found no evidence of return predictability using sentiment indicators, and of the information content of posting volume for forecasting volatility. Leung and Ton [3] collected 2.5 million messages from Hotcopper (the biggest Australian stock discussion forum). The messages were classified using NB with a manually classified training set of 10,000 messages. They found that the number of board messages and message sentiment significantly and positively relate to the contemporaneous returns of underperforming (low ROE, EBIT margin, EPS) small capitalization stocks with high market growth potential.

The goal of this paper is to overcome the limitation of previous studies. Prior studies have used varied machine learning classifiers, but no comprehensive comparison has been made between different classifiers. Also, the nature of microblogging (short in length, use of slangs and typo errors) calls for sophisticated pre-processing before the messages could be fed to machine learning algorithms. Finally, many metadata from messages could be used to increase the performance of these algorithms.

3   Data

We have focused on top ten US stocks based on market capitalization: Apple (AAPL), Alphabet (GOOG, GOOGL), Microsoft (MSFT), Amazon (AMZN), Berkshire Hathaway (BRK.A, BRK.B), Exxon Mobil (XOM), Facebook (FB), Johnson & Johnson (JNJ), General Electric (GE), Wells Fargo (WFC). For each stock, we have collected messages posted on StockTwits from January 01, 2016 to June 31, 2016. We have randomly selected 20,000 tweets for this research.

StockTwits (http://stocktwits.com/) was selected as our data source for this study. StockTwits is a social media platform designed for sharing ideas between various stakeholders, such as, investors, traders and entrepreneurs, etc., and it is a popular platform, which had 230,000 active users in June 2013. Messages are limited to 140 characters but may contain links, charts or even video, similar to Twitter. However, in contrast to Twitter, StockTwits only focuses on the stock market and stock investment, which makes it a less noisy data source than other general microblogging services, such as Twitter. Each message contains at least one $cashtag (i.e., $AAPL, $AMZN, $GOOG). Since September 2012, users are able to disclose their sentiment for each message (post) as "Bullish" or "Bearish". Since this data contains self-disclosed sentiments, it can be used to test machine-learning algorithms, without manual classification.

3.1 Pre-processing of Data

To remove noise from messages, We have pre-processed all the messages (Agarwal et al. [13]) as following: (1) replace all URLs with a tag ||U||, (2) replace all targets (e.g. "@Sam") and all cashtags (e.g. "$AAPL") with tag ||T|| (3) replace all negations (e.g. not, no, never, n't, cannot) with notation "NOT", and (4) replace a sequence of repeated characters by three characters, for instance, convert goooood to good.

Afterwards, we have processed the tweets using natural language processing tools: (1) use Stanford tokenizer [14] to tokenize the tweets. (2) use a port-of-speech tagger to process tokenized message and attach a part of speech tag to each word. (3) use the stopword list in Python NLTK to identify and remove stopwords from each message. (4) punctuations are also removed from messages. (5) Then we use WordNet [15] to find English words. (6) get the stem of each word using Porter stemmer.

3.2 Prior Polarity Scoring

We based some of our features on the prior polarity of words [13]. In this case, Dictionary of Affect in Language (DAL) is used and extended by WordNet. DAL contains about 8000 English words with a pleasantness score between 1 to -3 (negative to positive) for each word. We normalise the scores by dividing all the scores by 3. Words with polarity less than 0.5 are treated as negative, while words with polarity higher than 0.8 are treated as positive and the rest is treated as neutral. When a word is not found in the DAL dictionary, all synonyms are retrieved from WordNet. We then search for each of the synonyms in DAL. If any synonym is from DAL, the same pleasantness score of the original word in DAL is assigned to its synonym. If none of the synonyms appears in DAL, then the word is not linked with any prior polarity.

3.3 Features

Following Agarwal et al. [13], the features that we use could be divided into four classes: first, a list of words from the training set, and the occurrence of these words for each tweet as Boolean values. Second, counts of primary features, which result in a natural number ($\in$ N). Third, features whose value is a real number ($\in$ R). Fourth, features whose values are Boolean ($\in$ B). Each of these general classes is further divided into two subclasses: Polar features VS Non-polar features. We classify a feature as polar if we find it prior polarity by searching DAL (extended by WordNet). All the other features, which do not have any prior polarity fall in the Non-polar category. Finally, Each of Polar and Non-polar features are divided into two subclasses: POS and Other. POS is features which are parts-of-speech (POS) of words, with types of JJ (Adjective), RB (Adverb), VB (Verb), NN (Noun).

Same as Agarwal et al. [13], row f1 belongs to class Polar POS and is the count of the number of positive and negative POS in messages. f2, f3, f4 all belongs to class Polar Other. f2 is the number of negation words and positive and negative prior polarity. f3 is the number of ($\pm$) hashtags, capitalised words, and words with exclamation marks. f4 belongs to Non-polar POS and is the number of different part of speech tags. f5, f6 belong to Non-polar Other. f5 is other words without polarity; f6 is the number of hashtags, URLS, targets and cashtags. f7 belongs to Polar POS and is the sum of prior polarity scores of words with POS of JJ, RB, VB, and NN. f8 belongs to Polar Other and is the sum of prior polarity scores of all words. f9 refers to class Non-Polar Other and is the percentage of tweets that is capitalised. Finally, f10 belongs to class Non-Polar Other and is the presence of exclamation and presence of capitalised words. The descriptions are shown in Table 1.

Table 1. Summary statistics.

N	Polar	POS	# of ($\pm$) POS (JJ, RB, VB, NN)	f1
		Other	# of negation words, positive words, negative words	f2
			# of ($\pm$) hashtags, capitalised words, exclamation words	f3
	Non-polar	POS	# of POS (JJ, RB, VB, NN)	f4
		Other	# of words without prior polarity	f5
			# of hashtags, URLs, targets, cashtags	f6
P	Polar	POS	For POS, $\sum$ prior polarity score of words that POS	f7
		Other	$\sum$ prior polarity scores of all words	f8
	Non-polar	Other	Percentage of capitalised text	f9
B	Non-polar	Other	Exclamation, capitalised text	f10

4 Machine Learning Models

In this research, we use three different classifiers: Naïve Bayes (NB), Logistic Regression (LR), and Support Vector Machine (SVM). We choose these three classifiers, as NB and SVM are two most widely used classifiers in the social media sentiment analytics in a financial context and LR is a good approach for 2-way classification (classify dataset into two groups), while has not been explored in comparison with other two classifiers in the social media sentiment analytics in a financial context. Each classifier is tested using a 10-fold cross-validation, which is a common practice with machine-learning classifiers. For Naïve Bayes, we use Multinomial NB and Bernoulli NB. For SVM, we use three different kernels, which are linear, poly, and rbf kernels.

4.1 Naïve Bayes

NB is based on Bayes' theorem with the naïve assumption of independence between every pair of features. Given C stands for a class and W_1 to W_n are the feature vector, Bayes' theorem states the following:

$$P(C|W_1, \ldots, W_n) = \frac{P(C)P(W_1, \ldots, W_n|C)}{P(W_1, \ldots, W_n)} \tag{1}$$

The Naïve assumption gives that:

$$P(W_1, \ldots, W_n|C) = \prod_{i=1}^{n} P(W_i|C) \tag{2}$$

The relationship of Eq. 1 is then simplified to:

$$P(C|W_1, \ldots, W_n) = \frac{P(C) \prod_{i=1}^{n} P(W_i|C)}{P(W_1, \ldots, W_n)} \tag{3}$$

As $P(W_1, \ldots, W_n)$ is always a constant value given the input $(W_1$ to $W_n)$, we can apply the following classification rule:

$$P(C|W_1, \ldots, W_n) \propto P(C) \prod_{i=1}^{n} P(W_i|C) \tag{4}$$

Finally, the classification with the highest posterior probability is chosen.

$$\hat{C} = argmaxP(C) \prod_{i=1}^{n} P(W_i|C) \tag{5}$$

The main difference between NB classifiers is the assumptions that they make regarding the distribution of $P(W_i|C)$.

Multinomial NB uses the NB algorithm for multinomial distributed data. $P(W_i|C)$ is estimated by a smoothed version of maximum likelihood:

$$P(W_i|C) = \frac{N_{Ci} + \alpha}{N_C + \alpha n} \tag{6}$$

where N_{Ci} is total number of times feature W_i falls in a sample of class C in the training set, and N_C is the total number of all features for class C. α is the smoothing parameter and prevent zero probabilities.

Bernoulli NB uses the NB classifier for multivariate Bernoulli distributed data. The decision rule for Bernoulli NB is based on:

$$P(W_i|C) = P(i|C)W_i + (1 - P(i|C))(1 - W_i) \tag{7}$$

Which penalised the non-occurrence of a feature i that is an indicator for class C.

4.2 Logistic Regression

The logistic function $\sigma(t)$ is defined as follows:

$$\sigma(t) = \frac{1}{1 + e^{-t}} \tag{8}$$

Let's presume that t is a function of the independent variables $(W_1, \ldots, W_n)$, where:

$$t = f(W_1, \ldots, W_n) \tag{9}$$

And the logistic function could be written as:

$$F(W_1, \ldots, W_n) = \frac{1}{1 + e^{-f(W_1, \ldots, W_n)}} \tag{10}$$

$F(x)$ is described as the probability of the dependent variable (C) is a "Bullish" or "Bearish".

4.3 Support Vector Machine

SVMs are a group of supervised learning algorithms widely used for classification. To have an overview of SVMs, SVMs provide a separation boundary (linear or non-linear) in the dataset. Let us consider a training set with n observations (x_i). Each of the observations is a p-dimensional vector of features. Each training set has a self-disclosed label (y_i) in this research. Then a hyperplane or a hypersurface is constructed that could separate the training dataset with respect to the labels. To balance the problem of over-fitting and under-fitting, a parameter is introduced into the model: penalty parameter C of the error term. The lower your C value, the smoother and more generalised your decision boundary is going to be. But if you have a large C value, the classifier will attempt to do whatever is in its power to perfectly separate each sample to correctly classify it.

Kernels methods enable SVMs to be functional in a higher dimensional, implicit feature space, without calculating the coordinates of data in that space, but rather by calculating the inner products between all pairs of data.

4.4 Measures

We measure the accuracy, precision, recall and F1 measures for all the classifiers.

$$Accuracy = \frac{tp + tn}{tp + tn + fp + fn} \tag{11}$$

$$Precision = \frac{tp}{tp + fp} \tag{12}$$

$$Recall = \frac{tp}{tp + fn} \tag{13}$$

$$F1 = 2 * \frac{Precision * Recall}{Precision + Recall} \tag{14}$$

where tp is true positive, tn is true negative, fp is false positive, and fn is false negative.

5 Results

We use 20,000 messages for this research. Each message has a self-disclosed sentiment, which is "Bullish" or "Bearish". This provides good training sets as well testing
sets for the classifiers. We do a 10-cross validations for this study. The original dataset
is partitioned in 10 equal size subsamples. In the ten subsamples, a single subsample is
used as the testing dataset, while the rest nine subsamples are used as the training set.
Then we report the average accuracy, precision, recall, and F1 measure for all the
experiments with different size of data. Figure 1 shows the learning curve for the 2-way
classification. "MNB" is Multinomial NB, "BNB" is Bernoulli NB, "LR" is Logistic
Regression, "LSVC" is Linear SVM, "SVC_poly" is SVM using poly kernel, and
"SVC_rbf" is SVM using radial basis function kernel.

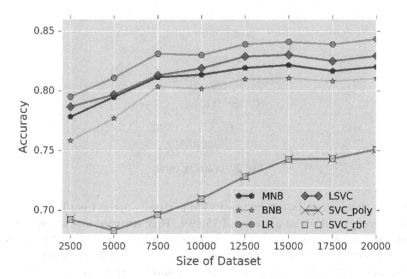

Fig. 1. Learning curve

It is clear that Logistic Regression Classifier outperforms all the other classifiers in
this 2-way classification. However, Logistic Regression is not widely used in the
classification of messages from social media in literature. In this case, we encourage
researchers to use more classifiers and compare the accuracy of the classifiers, instead
of only focusing on one or two classifiers with one kind of kernel. Figure 1 also shows
that there is a quite sharp increase in accuracy when the size of dataset moves over
7,500. Thus, we encourage researchers to use a training set of more than 7500 to have a
good accuracy in classification.

Overall accuracy, precision, recall and F-Measure are summarised in Fig. 2. There
is a trade-off between recall and precision. Thus researchers have used F-Measure to
determine which method is superior to others. Logistic Regression classifier has the
highest value for accuracy (0.844) and F-Measure (0.901). This means that LR

out-performs other classifiers in the social media sentiment analytics in a financial context. We also notice that SVMs with poly and rbf kernel have a recall of value 1 and the lowest precision among all the classifiers. This means that these two classifiers have no false negative classification and have a great amount of false positive classification, which makes these two classifiers have really poor performance.

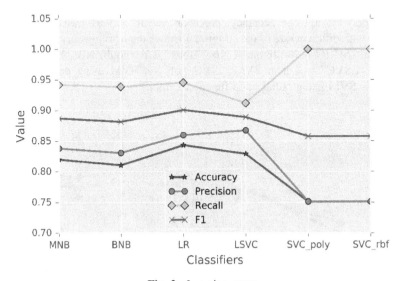

Fig. 2. Learning curve

Previous research on Twitter has used SVM to classify tweets from Twitter into two sentiment groups and got an accuracy of 75.39%. They streamed the data in real-time. No language, location or any other kind of restriction was made during the streaming process. Tweets in foreign languages are converted it into English using Google translate before the annotation process. They manually annotated 11,875 tweets. In comparison, our research comes up with an accuracy of 81.9% using Linear SVM, with a dataset of 10,000 tweets. Using almost the same method (unigram and metadata features), the accuracy for StockTwits outperform Twitter. The reasons could be: (1) StocksTwits is focusing on the financial market, which has less noise. (2) There is a great portion of users in StockTwits who are investors or traders. These people use more formal and accurate words than average users in Twitter. In this case, StockTwits is considered as a better data source to conduct sentiment analysis, especially in a financial context.

6   Conclusion

In this study, we achieve the following: First, we find that among the three classifiers, Logistic Regression performs the best in classifying messages on StockTwits. Though prior research studies analyzing financial microblogging services have been using NB or SVM, we report a superior performance of Logistic Regression in this environment. Second, we get a better accuracy using messages from StockTwits than from Twitter as a data source. When we want to find the correlation between social media sentiment and stock market variables, we want to include as many messages from social media platforms as we could. This gives rise to the need to classify all messages (with or without a self-disclosed sentiment) from a social media platform. Thus, we posit that StockTwits could be a better data source than Twitter to analyze sentiments in financial markets.

References

1. Chen, H., De, P., Hwang, B.-H.: Wisdom of crowds: the value of stock opinions transmitted through social media. Rev. Financ. Stud. **27**, 1367–1403 (2014). doi:10.1093/rfs/hhu001
2. Oh, C., Sheng, O.: Investigating predictive power of stock micro blog sentiment in forecasting future stock price directional movement (2011)
3. Leung, H., Ton, T.: The impact of internet stock message boards on cross-sectional returns of small-capitalization stocks. J. Bank. Financ. **55**, 37–55 (2015). doi:10.1016/j.jbankfin. 2015.01.009
4. Bollen, J., Mao, H., Zeng, X.: Twitter mood predicts the stock market. J. Comput. Sci. **2**, 1–8 (2011). doi:10.1016/j.jocs.2010.12.007
5. Loughran, T., McDonald, B.: Textual analysis in finance and accounting: a survey. SSRN Electron. J. (2015). doi:10.2139/ssrn.2504147
6. Henry, E.: Are investors influenced by how earnings press releases are written? J. Bus. Commun. **45**, 363–407 (2008). doi:10.1177/0021943608319388
7. Loughran, T., Mcdonald, B.: When is a liability not a liability? Textual analysis, dictionaries, and 10-Ks. J. Financ. **66**, 35–65 (2011). doi:10.1111/j.1540-6261.2010.01625.x
8. Hu, T., Tripathi, A.K.: The performance evaluation of textual analysis tools in financial markets (2015)
9. Antweiler, W., Frank, M.Z.: Is all that talk just noise? The information content of internet stock message boards. J. Financ. **59**, 1259–1294 (2004). doi:10.1111/j.1540-6261.2004. 00662.x
10. Sprenger, T.O., Tumasjan, A., Sandner, P.G., Welpe, I.M.: Tweets and trades: the information content of stock microblogs. Eur. Financ. Manag. **20**, 926–957 (2014)
11. Tirunillai, S., Tellis, G.J.: Does chatter really matter? Dynamics of user-generated content and stock performance. Mark. Sci. **31**, 198–215 (2012). doi:10.1287/mksc.1110.0682
12. Oliveira, N., Cortez, P., Areal, N.: On the predictability of stock market behavior using stocktwits sentiment and posting volume. In: Correia, L., Reis, L.P., Cascalho, J. (eds.) EPIA 2013. LNCS, vol. 8154, pp. 355–365. Springer, Heidelberg (2013). doi:10.1007/978-3-642-40669-0_31
13. Agarwal, A., Xie, B., Vovsha, I., Rambow, O., Passonneau, R.: Sentiment analysis of Twitter data. In: Proceedings of the Workshop on Languages in Social Media, pp. 30–38. Association for Computational Linguistics (2011)
14. Klein, D., Manning, C.D.: Accurate unlexicalized parsing. In: Proceedings of the 41st Annual Meeting on Association for Computational Linguistics-Volume 1, pp. 423–430. Association for Computational Linguistics (2003)
15. Miller, G., Fellbaum, C.: Wordnet: An electronic lexical database (1998)

The Cannibalization Effects of New Product Preannouncement and Launch on the C2C Marketplace

Dan Ke[1(✉)], Heci Zhang[1], and Yanbin Tu[2]

[1] Wuhan University, Wuhan, China
dkeuconn@gmail.com, 492071661@qq.com
[2] Robert Morris University, Moon, USA
tu@rmu.edu

Abstract. This paper aims to investigate whether and how the introduction of a new product impacts the consumer behaviors on the C2C marketplace. We take iPhone preannouncement and launch as a case to examine whether there is the cannibalization effect on existing Apple iPod product and Microsoft Zune on eBay. Difference-in-Differences (DID) Model is applied as the main econometrical analysis methods in this empirical study. We collected all the bidding records from December 10th 2006 to August 20th 2007 (including the date of iPhone preannouncement January 9th 2007 and its launching into the market June 29th 2007) for DID analysis. Our analysis results show that new product introduction cannibalizes the other existing products of the same firm; the preannouncement and the launching of new product both promote the occurrence of cannibalization; specifically, new product cannibalizes the most recent version of existing products more significantly than others.

Keywords: Difference-in-differences model · Cannibalization · New product preannouncement · New product launch · C2C marketplace

1 Introduction

Online shopping has been the most popular way in daily purchase for consumers due to the significant reduction of search costs (Bakos 1997). Electronic auction, especially online C2C auction market, has exerted a significance influence on people's purchase behavior by its unique way. Online C2C platforms, such as eBay, Amazon, have gained extensive attention from researchers. Although much research about the online auction market contributes to this research field, there has been little effort on the effects of the introduction of new product on the online auction market. In this study, we are interested in explore whether and how a new product introduction would influence consumer behaviors of online C2C marketplace.

Innovation, the process of bringing new products to market, is one of the most important issues for firms and researchers alike (Hauser et al. 2006). We take the innovative iPhone's preannouncement and launch as a case to examine the influence of new product to the existing products at online C2C marketplace in our paper. On January 9th 2007, Steve Jobs announced on the launch event that Apple is introducing

M. Fan et al. (Eds.): WeB 2016, LNBIP 296, pp. 84–94, 2017.
https://doi.org/10.1007/978-3-319-69644-7_8

a new revolutionary product including three separate devices such as a widescreen iPod with touch controls. It was the earliest version what we call iPhone. The appearance of iPhone changed the whole mobile communication industry, and led to huge responses among consumers.

Specifically, the research questions of this paper are: How does the introduction, the preannouncement and official launch, of a new product impacts the consumer behaviors on existing product on the C2C marketplace? How does the new product introduction impact the products from the same firm and from the rival firm differently?

2 Related Literature and Research Hypotheses

2.1 Literature Review

New Product Preannouncement

The original research about the preannouncement of new product starts from the signal theory in economics, and many researchers held that preannouncement is one form of market signals. Porter (1980) classified the market signals into two types: preannouncements of market actions and market actions themselves. Preannouncements is intended to convey information or to gain information from competitors, defined as competitive signal to declare the intention and commitment of the new product of firm. Preannouncements may be directed to one or more audiences, such as customers or end consumers, competitors, shareholders, or distributors (Heil and Robertson 1991). Eliashberg and Robertson (1988) and Robertson (1995) defined "preannouncement" as a formal, deliberate communication before a firm actually undertakes a particular marketing action such as a price change, a new advertising campaign, or a product line change. The preannouncing behavior pertains to new products or services (excluding flankers or line extensions). They suggested that preannouncement behavior is motivated when the new products involve higher customer switching cost, the preannouncing firms have no market dominance, preannouncing firms are smaller, and the competitive environment is less combative (Thorbjornsen et al. 2016). Robin and Moore (1989) suggested that the purpose of preannouncement is to arouse the consumers' curiosities and interest and then provide demand motivation.

New product introduction may create opportunities for product differentiation and competitive advantage and can have a positive impact on the firm's value that preannounced a new product (Kleinschmidt and Cooper 1991). New product introductions have positive effects on the market value of the announcing firms (Chen and Ho 1997). For many firms, it is beneficial to communicate their development activities to internal and external audiences in advance of a new product introduction (Lilly and Waiters 1997). Announcing the future availability of new products is widely practiced (e.g., Singh 1997). The value of a new product preannouncement was the greatest for the most technologically based industries such as computers and electrical equipment and appliances. Thus, continued emphasis on new products should positively affect the value of firms in these industries (Chaney et al. 1991). Researchers always focused on the introduction effect on competitors because many firms take advantage of new product preannouncement as a competition strategy against rivals even some firms utilize it just as a "vaporware" intentionally to mislead and confuse their opponents (Bayus et al. 2001). Mahajan et al. (1993) found that the effects of new product pre-announcements on the competitors depend on the interaction of the following two effects: the market expansion effect and the market substitution effect. The competitive and aggressive reaction of incumbent depends on the perceived hostility and credibility of new product preannouncement signal (Robertson et al. 2005).

Cannibalization Effect

Past researchers had paid little attention to the impacts of new product preannounce-ment on earlier versions or other existing products of the same firm. Instead, it has been widely discussed that the launch of new product significantly affects the price and sales of the existing products of the same firm when the consumers become aware of it. This phenomenon is called as cannibalization. This study is investigating whether the pre-announcement of new product has the analogous effect of cannibalization on existing products on the C2C marketplace before it has officially launched on the market.

In general, the effect of cannibalization means the competition among the products existing in a same market segments. Actually, cannibalization is an extremely impor-tant concept in marketing, which has attracted wide attention of researchers on mar-keting. Copulsky (1976) first proposed the conception: Cannibalism in the Marketplace by analyzing the competition among General Foods's new products and older products. Mazumdar et al. (1996) suggested that the cannibalization occurs when two or more products of a firm compete and take away each other's sales in a given product-market, especially when the firm is pushing a new product into the market. Cannibalization refers to a reduction in the market share of one product as a result of the introduction of a new product by the same producer (Cheng and Peng 2012).

In today's markets, almost all firms are affected by cannibalization because a majority of new products are minor modifications or line extensions of existing products (Mason and Milne 1994). Cannibalization was typically viewed as a phe-nomenon having an adverse impact on corporate performance (Traylor 1986). Yet, many firms routinely and consciously develop and introduce new products that can-nibalize existing products (Mazumdar et al. 1996).

In early literature on cannibalization effect, researchers who studied this phe-nomenon developed a model that considers the impact on the incremental costs and the incremental revenues from line extensions (Ramdas and Sawhney 2001). Ramdas and Sawhney (2001) pointed out that cannibalization is estimated indirectly by first esti-mating consumer utility for a new product via part-worth utility functions over a few product attributes, and then using these to estimate cannibalization. This indirect approach has limitations when some product dimensions, such as aesthetics and so on, cannot be adequately described by a few objective attributes (Srinivasan et al. 1997). To overcome this problem they measured pair-wise cannibalization between prototypes directly, without relying on a decompositional approach to estimate consumer utility. This procedure was simpler than the conjoint measurement procedure, and was more valid for products with a high aesthetic dimension. Dahan and Srinivasan (1998) used prototypes in conjoint concept selection, but they do not incorporate costs.

Prior work of new product introductions used individual consumer attitudinal or behavioral data to estimate the cannibalization effect on existing product sales (Urban and Hauser 1993). Diffusion models was used to estimate the extent of cannibalization due to the firm's or the competitors' product introductions (Bass 1969). Peterson and Mahajan (1978), Sharma and Buzzell (1993) examined the simultaneous diffusion of multiple innovations, in which the diffusion of a later innovation can substitute for the diffusion of an earlier one. Norton and Bass (1987), extending the Bass model, examined the displacement of preceding generations of a high-technology product (complete cannibalization) by a new generation of the existing products.

Since few published work exists to determine empirically cannibalization effects from aggregate sales data, a conceptual framework for assessing an extension's cannibalization or expansion of parent brand sales was presented by Reddy et al. (1994). A simple approach was the use of a dummy variable regression, which would provide an estimate of the effect as a constant shift in the intercept term. Another approach was to use intervention analysis (Box and Tiao 1975). The advantage of intervention analysis over a dummy variable method is that one could specify nonlinear effects of the intervention as opposed to simple step function.

With the deepening of the study on the question, the empirical literature of marketing on cannibalization effect of new product launch is particularly rich. Heerde et al. (2010) aimed at the cannibalization effects both within focal category and across categories based on the data from automotive industry. Some researchers separated the cannibalization effect and market stealing effect and set them in opposite. Haynes et al. (2014) suggested that the cannibalization effect is quite obvious for an own product introduction, particularly if the newcomer exhibits vertical superiority over the existing product, while the largest market stealing impact is from the introduction of newer rival brands.

From another perspective, Samiee et al. (2014) explored a new form of cannibalization as a strategic choice named intentional cannibalization, considering that the deliberate cannibalization of products in favor of new innovations is a subtle marketing practice that is common among firms that want to combat existing or potential competitive threat.

However, few empirical literature focus on the relationship between new product preannouncement and the cannibalization effect, although the researchers studying the preannouncement effect realized that the new product preannouncement may lead to the cannibalization and affect the existing products in the market long time ago. Some research pointed out that the new product preannouncement will carry the risk of cannibalization, after the preannouncement, firms' existing products are no longer considered new, and some customers choose to wait for the future product. This purchase-freezing behavior may potentially reduce the sales in the product category. The larger the preannouncing firm's existing market share is, the higher the cannibalization will be. Risks are costly in most signaling behavior (Eliashberg and Robertson 1988; Su and Rao 2007). Wu et al. (2004) suggested that the firms introduce new product beyond preannounced deadlines due to some factors, the increased delay in introduction of preannounced new product is positively associated with likelihood of cannibalization.

2.2 Research Hypotheses

In this study, we focus on its influence brought on the online consumer behavior. Lilly and Walter (1997) pointed out that preannouncement can delay recent purchases to wait for the new product available. Once upon the preannouncement of new product, the consumers who have an intention to update their product will not spend too much resource and money to the existing products on the market. Due to the reduction of the switching cost, the entry barrier is about to be eliminated. At the meantime, the preannouncement urges the consumers to search information of new product and its complements and reserve the budget for the new version (Lilly and Waiters 1997). Preannouncement may harm the sales of the firm's existing products, commonly referred to as the cannibalization effect. Some current customers may choose to stop purchasing the existing product and wait for the future product, reflected in either the pent-up demand or the prelaunch order for the new product (Su and Rao 2010).

Based on the sufficient and supporting theories above, we propose the following hypotheses rationally and logically: The new product preannouncement and launch leads to the cannibalization effects on the existing product on the C2C marketplace.

H1: The preannouncement and launch of new product lead to decreased winning price of products from the same firm and products from the rival firm on the C2C marketplace.

H2A: The preannouncement of new product impacts the product from the same firm more significantly, i.e., the winning price of the product from the same firm decreased more after the new product preannouncement.

H2B: The launch of new product impacts the product from the same firm more significantly, i.e., the winning price of the product from the rival firm decreased more after the new product launch.

3 Research Model and Data Collection

We apply Difference-in-differences (DID) model to analyze effects of new product preannouncement and new product launch. We use Apple iPhone as the context and examine how its preannouncement and launch influence other existing Apple products and its complementary product on the market.

DID is widely used to evaluate the impact of policy. Researchers usually measure the treatment effects such as changes in medical research by DID. It is also popular in evaluating the result of policy or public interventions in economics (Qin and Zhang 2008). In marketing research, DID could be applied to evaluate the impact of market actions such as new product preannouncement, new product launch, etc. We applied DID model to explore whether the cannibalization effect exists between iPhone and iPod and treated the preannouncement event as a policy.

In DID estimation, the change in outcomes in the treated group pre- and post-treatment is compared with the change in outcomes over the same time period for a control group which did not receive the treatment. In this study, we regarded the introduction of a new breakthrough product iPhone as a strong treatment. We estimated the impact of the introduction of iPhone on differences in the price in control group and the treated group before and after the introduction of the new product.

3.1   Data Set

Suitable data sets for the estimation of the model specified above require a number of features. First, they must have a panel or matched cross-section element of at least two time periods: the model estimates the effect of cannibalization as a function of the introduction at time t^*. Second, they must provide information of the sales, such as the winning bid price of the product and auction ending date and time. Third, they need to be part of a series: there must be observations for which the time interval straddles the introduction of the new product on June 26 2007. Fourth, they must provide information on other factors that might influence the effect, such as the shipping cost, duration of the auction and the number of bids attracted and so on. Finally, they must provide reasonably large samples of individuals.

The construction of the DID model needs adequate cell sizes of individuals. We obtained the appropriate data from eBay platform, including enough auction information of many kinds of digital product from December 10th 2006 to August 20th 2007 including the date of new product preannouncement (January 9th 2007) and launching into the market (June 29th 2007).

To test the cannibalization effect working to the products from the same firm and from the rival firm, we selected Microsoft Zune and Apple iPod Nano2. The two devices both are MP3 player with same functions and similar prices so that they are suitable to be selected as the research objects.

3.2   Econometric Model

The DID estimator controls both the group-specific and time-specific effects (Blundell and Dias 2001). This approach is only possible because of the quasi-experiment in which the treated group is affected by the launch, whereas the control group is not. For example, if we only had the bidding information of the treated group before and after the introduction, we would not determine whether any change results from of the new introduction or from other reasons; similarly, if we only had the information of after the change, we would not be able to tell whether any difference between the treated and control group was due to the introduction or due to some other unobserved difference. Since we had access to the auction data before and after the introduction of the new product and in two products, we were able to estimate the impact of the cannibalization alone.

Suppose that the effect is introduced at a point in time t^*, and that for observations prior to t^* no new product is in the market, define t is a time variable if the goods was sold before t^* taking the value 1 (if after t^* taking 2). Considering two groups of individuals, suppose those in group $G_t = 1$ (if i belongs to the "treatment" group) are directly affected by iPhone release, while those which are 2 GB in group $G_t = 0$ (if i is belong to the "control" group) are not affected by iPhone introduction. The following linear model is specified for the winning bid price:

.

$$y_{it} = \beta_0 + \beta_1 G_i \cdot D_t + \beta_2 D_t + \beta_3 G_i + \varepsilon_{it}$$

where β_0 is constant term, G_t is a dummy variable for the treatment group (or the control group), β_2 shows how the price of the product changes if there is no introduction of new product, β_3 is used to measure the difference between control group and treatment group without affecting by time effect and D_t is a dummy variable taking the value 1 (if t = 2, if not taking the value 0), the interaction term $G_t \cdot D_t$ is the effect we need. This is a simple difference-in-differences estimator given by double differencing model. Consider the case of two time periods: before and after the release of iPhone. The OLS estimator of β_1 is given by differencing across these two groups and two time periods: $\{[y_{treat,2} - y_{treat,1}] - [y_{control,2} - y_{control,1}]\}$, where y_{it} is the winning bid price on eBay.

This specification assumes that in the absence of the new product the difference in price between the two groups is the same in each time period, or equivalently that the changing tendency of price over time is the same for each group. This is the first key identifying assumption and will be returned to below. The potential problem with this assumption is that even in the absence of the new iPhone release, the changing tendency of price may evolve differently in the different groups. The second key identifying assumption is that the release does not alter the price in the control group $(G_t = 0)$.

In order to estimate the cannibalization effect of the new introduction, we utilized the adjusted difference-in-differences estimate, so the model could be extended to:

$$y_{it} = \beta_0 + \beta_1 G_i \cdot D_t + \beta_2 D_t + \beta_3 G_i + \gamma_1 duration + \gamma_2 nobi + \gamma_3 pofb + \varepsilon_{it} shco + \varepsilon_{it}$$

Where the duration is the time between the beginning and the ending of one auction, the nobi is the number of bids attracted of the product, the pofb is the percentage of seller's positive feedback and the shco is the Shipping cost.

4 Empirical Results

In order to test how the new product introduction impacts the products from the same firm and from the rival firm differently. We used the same method and step to analyze the cannibalization effect of the new product preannouncement and launch. We selected two kinds of product-Zune made by Microsoft and iPod Nano2 made by Apple expecting to find some new findings. We defined the iPod Nano2 as the treated group and the Microsoft Zune as the control group (Table 1).

From Table 2 we observed the first difference for these two groups on the two stages respectively. We found the change of winning price for the Zune are $-17.34 on the preannouncement stage. The corresponding first differences for the iPod Nano2 are $-21.78. This shows that the average winning price declined more for iPod Nano2 because of the influence of new product preannouncement. But on the launch stage, a different and interesting results are presented, the first difference of the winning price of the Zune is $-3.72, while the iPod Nano2 almost no change. It seems to indicate that Zune was affected more seriously because of the new product launch into the market. As for the authenticity of this result, we need to test the DID model results.

Table 1. Descriptive statistics of iPod Nano2&Zune on the two stages

Stage	Product	Before/after	Observations	MEAN	MAX	MIN	SD	VAR
Preannouncement (Dec/10/2006– May/28/2007)	iPod Nano2	Before	616	133.7759	214.95	85	15.4327	238.1683
		After	2346	112.0048	350	24.99	19.17956	367.8557
	Zune	Before	961	211.2236	329	160	16.83332	283.3605
		after	3177	193.8815	310	128.05	16.35455	267.4714
Launch (May/10/2007– Jul/29/2007)	iPod Nano2	Before	1213	110.0446	289	45	20.182	407.3133
		After	723	109.9511	259.95	51	21.00603	441.2531
	Zune	Before	1047	185.7309	305	136.59	19.05204	362.9804
		After	751	182.0068	360	75	22.95029	526.7159

Table 2. Mean values and first difference of iPod Nano2&Zune on the two stages

	Before preannouncement	After preannouncement	First difference
Control (Zune)	211.22	193.88	−17.34
Treated (Nano2)	133.78	112.00	−21.78
	Before launch	After launch	First difference
Control (Zune)	185.73	182.01	−3.72
Treated (Nano2)	110.04	109.95	−0.09

Table 3 presents the results for the cannibalization effect of new product between the product from the same firm and the product from the rival firm. The key coefficient β_1 (−3.279***) is significant and negative on the preannouncement, indicating that there was an overall decrease in price in iPod Nano2 (the treated group) following these months as compared to the Zune (the control group).

This decrease in price shows the effect of new product preannouncement on the product from the same firm is greater. But on the launch stage, the coefficient β_1 is significant and positive, indicating that the product from the rival firm decreased more

Table 3. The DID outcomes of iPod Nano2&Zune

Variables	Preannouncement coefficient (standard error)	Launch coefficient (standard error)
Dt*Gi (β_1)	−3.279***(0.983)	3.115**(1.361)
Dt (β_2)	−18.55***(0.623)	−4.174***(1.012)
Gi (β_3)	−77.70***(0.848)	−74.65***(0.889)
Duration	0.698***(0.0913)	0.455***(0.131)
Nobi	−0.249***(0.0235)	−0.359***(0.0338)
Pofb	0.339***(0.0962)	0.394***(0.101)
Shco	−0.352***(0.0752)	0.188***(0.0528)
Observations	7,005	3,441
R-squared	0.858	0.789

Robust standard errors in parentheses
***p < 0.01, **p < 0.05, *p < 0.1

compared to the product from the same firm due to the new product launch. We found that the coefficient β_2 is significant and negative as well, suggesting that if there was no new product introduction, how the price of the products changes. The significant and positive coefficient β_3 acknowledged us that there was significant difference between the control group and the treated group for the winning price. The result presented to us shows that the cannibalization effect of new product launching into the market is exist. The result of the duration, *pofb* and *shco* is same as the first experiment. But the *nobi* which is negative shows that the number of bids attracted affected the winning price negatively. The result presented to us shows that the cannibalization effect of new product preannouncement and launch is exist and the effect on the different stages are different between the product from the same firm and from the rival firm. On the preannouncement, the iPod Nano2 was cannibalized greater by the preannouncement of iPhone. But on the launch stage the Zune was affected more.

5   Conclusions

In this paper, we investigated the effects of new product cannibalization, in the form of preannouncement and launching into the market. In doing so, we provided new contributions to understanding how the new product cannibalizes the other existing product which belong to the same firm. The result showed two effects: the preannouncement and the launch of new product both promote the occurrence of cannibalization.

Difference in differences model was used to analyze the effect of cannibalization working to the different firm's products. On the preannouncement stage, the significant coefficient indicates that the preannouncement of iPhone decreased the price of iPod more than the Microsoft Zune. The reduction induced by the iPhone have important operational consequence, mainly arising from the consumers who are enthusiastic about apple's product transferred their target to the new product and forgot the previous product existing in the market. But on the launch stage, we get the completely opposite result, the negative coefficient shows that the Microsoft Zune decreased more than the iPod Nano2. When the new product launched into the market and then consumer could buy the physical good from apple stores. The high-quality iPhone brought consumers enjoyable experience, meanwhile enhancing the firm's reputation and word of mouth and then promoting the consumer loyalty to Apple. The good brand image alleviated the decreasing trend of iPod so that the Microsoft Zune was defeated at the corporate level entirely.

In addition, we provided new insights into the mechanism underlying these findings. Under the assumption that consumers who love the brand replace the manufacturer's previous product when she develops a new product, no matter when the new product is just announced or sold in the market, and that customers change their purchase decision as they get new information of product, we developed and tested a model to explain the observed patterns of demand. The model emphasized the informational role of the introduction; in particular, their ability to provide visceral information and to deliver either a full or more limited range of products that can be sampled.

References

Bayus, B.L., Jain, S., Rao, A.G.: Truth or consequences: an analysis of vaporware and new product announcements. J. Market. Res. **38**(1), 3–13 (2001)

Bass, F.M.: A new product growth model for consumer durables. Manag. Sci. **15**(January), 215–227 (1969)

Blundell, R., Dias, M.C., Meghir, C., Van Reenen, J.: Evaluating the Employment Impact of a Mandatory Job Search Assistance Program. Working Paper 01/20, IFS, University College London (2001)

Chen, S., Ho, K.: Market response to product-strategy and capital-expenditure announcements in Singapore: investment opportunities and free cash flow. Financ. Manag. **26**, 82–88 (1997)

Chaney, P.K., Devinney, T.M., Winer, R.S.: The impact of new product introductions on the market value of firms. J. Bus. **64**(October), 573–610 (1991)

Dahan, E., Srinivasan, V.S.: The predictive power of Internet based product concept testing using visual depiction and animation. Working paper, MIT Sloan School, Cambridge, MA (1998)

Eliashberg, J., Robertson, T.S.: New product preannouncing behavior: a market signaling study. J. Market. Res. **25**(3), 282–292 (1988)

Goldfarb, A., Tucker, C.: Conducting Research with Quasi-Experiments: A Guide for Marketers, 28 March 2014

Hauser, J.R., Tellis, G.J., Griffin, A.: Research on innovation: a review and agenda for marketing science. Market. Sci. **25**(6), 687–717 (2006)

Homburg, C., Bornemann, T., Totzek, D.: Preannouncing pioneering versus follower products: what should the message be? J. Acad. Market. Sci. **37**(3), 310–327 (2009)

Thorbjornsen, H., Dahlen, M., Lee, Y.H.: The effect of new product preannouncements on the evaluation of other brand products. J. Prod. Innov. Manag. **33**(3), 342 (2016). 14 p.

Qin, J., Zhang, B.: Empirical-likelihood-based difference-in-differences estimators. J. R. Stat. Soc. Ser. B (Stat. Methodol.) **70**(2), 329–349 (2008)

Bakos, J.Y.: Reducing buyer search costs: implications for electronic marketplaces. Manag. Sci. **43**(12), 1676–1692 (1997). Frontier Research on Information Systems and Economics

Ramdas, K., Sawhney, M.S.: A cross-functional approach to evaluating multiple line extensions for assembled products. Manag. Sci. **47**(1), 22–36 (2001). Design and Development

Kleinschmidt, E., Cooper, R.: The impact of product innovativeness on performance. J. Prod. Innov. Manag. **8**, 240–251 (1991)

Lilly, B., Walters, R.: Toward a model of new product preannouncement timing. J. Prod. Innov. Manag. **14**(January), 4–20 (1997)

Mahajan, V., Sharma, S., Buzzell, R.: Assessing the impact of competitive entry on market expansion and incumbent sales. J. Market. **57**, 39–52 (1993)

Haynes, M., Thompson, S., Wright, P.W.: New model introductions, cannibalization and market stealing: evidence from shopbot data. Manch. Sch. **82**(4), 385–408 (2014). 24 p. ISSN 14636786

Mason, C.H., Milne, G.R.: An approach for identifying cannibalization within product line extensions and multi-brand strategies. J. Bus. Res. **31**(2/3), 163 (1994)

Norton, J.A., Bass, F.M.: A diffusion theory model of adoption and substitution for successive generations of high technology products. Manag. Sci. **33**(September), 1069–1086 (1987)

Singh, J.: The Vaporware Game. CNet News, 25 April 1997. Accessed June 1997

Heil, O., Robertson, T.S.: Toward a theory of competitive market signaling: a research agenda. Strat. Manag. J. **12**(6), 403–418 (1991)

Porter, M.E.: Competitive Strategy: Techniques for Analyzing Industries and Competitors. Free Press, New York (1980)

Vragov, R.: Why is eBay the king of internet auctions? An institutional analysis. e-Serv. J. **3**(3), 5–28 (2004, 2005)

Srinivasan, V., Lovejoy, W.S., Beach, D.: Integrated product design for marketability and manufacturing. J. Market. Res. **34**(February), 154–163 (1997)

Hauser, J.R.: Design and Marketing of New Products. Prentice-Hall Inc., Englewood Cliffs (1993)

Peterson, R.A., Mahajan, V.: Multi product growth models. In: Sheth, J. (ed.) Research in Marketing. JAI Press, Greenwich (1978)

Rabino, S., Moore, T.E.: Managing new product announcements in the computer industry. Ind. Market. Manag. **18**(February), 35–43 (1989)

Sharma, S., Buzzell, R.D.: Assessing the impact of competitive entry on market expansion and incumbent sales. J. Market. **57**(July), 39–52 (1993)

Samiee, S., Sääksjärvi, M., Harmacioglu, N., Hultink, E.J.: Performance consequences of intentional cannibalization and radical innovations in Chinese and western enterprises. In: AMA Winter Educators' Conference Proceedings, vol. 25, pp. F-4–F-5 (2014). 2 p.

Reddy, S.K., Holak, S.L., Bhat, S.: To extend or not to extend: success determinants of line extensions. J. Market. Res. **31**(2), 243–262 (1994). Special Issue on Brand Management

Seamans, R., Zhu, F.: Responses to entry in multi-sided markets: the impact of craigslist on local newspapers. Manag. Sci. **60**(2), 476–493 (2014)

Su, M., Rao, V.R.: New product preannouncements as a signaling strategy: An audience-specific review and analysis. J. Prod. Innov. Manag. **27**, 658–672 (2010)

Mazumdar, T., Sivakumar, K., Wilemon, D.: Launching new products with cannibalization potential: an optimal timing framework. J. Market. Theory Pract. **4**(4), 83–93 (1996)

Traylor, M.B.: Cannibalism in multi brand firms. J. Consum. Market. **3**, 69–75 (1986)

Urban, G.L., Carter, T., Gaskin, S., Mucha, Z.: Market share rewards to pioneering brands: an empirical analysis and strategic implications. Manag. Sci. **32**(June), 645–659 (1986)

Copulsky, W.: Cannibalism in the marketplace. J. Market. **40**(4), 103–105 (1976)

Wu, Y., Balasubramanian, S., Mahajan, V.: When is a preannounced new product likely to be delayed? J. Market. **68**(2), 101–113 (2004)

Cheng, Y.-L., Peng, S.-K.: Quality and quantity competition in a multiproduct duopoly. South. Econ. J. **79**(1), 180–202 (2012)

Electronic Word of Behavior: Conceptual Framework and Research Design for Analyzing the Effect of Increased Digital Observability of Consumer Behaviors in a Movie Streaming Context

Katrine Kunst[1(⊠)], Ravi Vatrapu[1,2], and Abid Hussain[1]

[1] Department of Digitalization, Centre for Business Data Analytics,
Copenhagen Business School, Howitzvej 60, 4th floor,
2000 Frederiksberg, Denmark
{kk.digi, rv.digi, ah.digi}@cbs.dk,
katrine.kunst@gmail.com
[2] Westerdals - Oslo School of Arts,
Communication & Technology, Oslo, Norway

Abstract. In this research in progress-paper, we introduce the notion of 'Electronic Word of Behavior' (eWOB) to describe the phenomenon of consumers' product-related behaviors increasingly made observable by online social environments. We employ Observational Learning theory to conceptualize the notion of eWOB and generate hypotheses about how consumers influence each other by means of behavior in online social environments. We present a conceptual framework for categorizing eWOB, and propose a novel research design for a randomized controlled field experiment. Specifically, the ongoing experiment aims to analyze how the presence of individual-specific behavior-based social information in a movie streaming service affects potential users' attitude towards and intentions to use the service.

Keywords: Observational Learning · Electronic Word of Mouth · Electronic Word of Behavior · Social media · Social influence · Facebook · e-Commerce · e-Business

1 Introduction

The internet and the following digitization of a diverse range of products and services (i.e. books, music, hotel booking etc.), has significantly increased the observability of other consumers' behaviors [7, 16, 17]. Let's consider music. The move from having a physical CD collection in one's home to having a digital collection, open for friends and connected others to see on the music service Spotify, has vastly expanded the potential audience for individual consumers' consumption behaviors within the music domain. Further, digitization enables that not only is the collection observable to social

---

The original version of this chapter was revised: The title was incompletely shown. This has now been corrected. The erratum to this chapter is available at https://doi.org/10.1007/978-3-319-69644-7_24

© Springer International Publishing AG 2017
M. Fan et al. (Eds.): WeB 2016, LNBIP 296, pp. 95–103, 2017.
https://doi.org/10.1007/978-3-319-69644-7_9

others, but also the more detailed information about consumers' continuous product use behaviors such as which albums, artists, songs etc. are most listened to (i.e. most used), and what's being listened to right now. Taken together, digitization enables increased observability of not only consumers' choices but also for entirely new levels of observability of consumption behaviors.

Observational Learning (OL) theory [2] predicts that humans can influence each other by means of behavior. Consequently, usage of products which are easily observable (e.g. clothes, cell phones), are generally more likely to influence potential users than products whose use is less observable (e.g. deodorant, shampoo, etc.) [2, 20]. Against this backdrop, the possibility for increasing the observability of products with hitherto low inherent observability could have major organizational implications and gives rise to new research questions. Not surprisingly, a growing body of academic literature across disciplines such as information systems [14, 17, 21], marketing [1, 7, 22] management, and economics [4] have begun investigating how the disclosure of the past actions of others (i.e. behavior-based information) affects consumers' future decisions and usage of products.

However, there exists a conceptual as well as an empirical gap in extant research. The description of what's being analyzed is often diffuse in its wording, and not consistent across the literature. Oftentimes, the phenomenon of interest is referred to as "*observational learning information*" [7], "*action-based social information*" [8], or simply "*product information*" or "*information about other online users' choices and product popularity*" [11]. Consequently, in this research we introduce the concept of 'Electronic Word of Behavior' (eWOB) defined as "*observable, online traces of consumer behavior*", and offer a conceptual framework to categorize different kinds of eWOB. On the empirical side, the majority of extant research analyzes the impact of either (a) aggregated behavior-based information incorporated into a product or service (i.e. the impact on sales when presented with information that "X no. of people have bought this" [11] or (b) when individual-specific behavior-based information is disclosed on e.g. Facebook (i.e. messages shared to Facebook that "John just voted" [5] or "John just reached this level in Game X" [1]. Consequently, there exists a research gap when it comes to study of the impact of incorporating *individual-specific* behavior-based information - especially from friends - *into* a product or service. In addition to our conceptual contribution to extant knowledge, we set out to empirically investigate the impact of incorporating individual-specific behavior-based information into a movie streaming service.

The rest of the paper is organized as follows. First we provide a brief exposition of the theory of OL that informs the notion of eWOB. Second, we generate hypotheses to test and compare the effect of behaviors (eWOB) vs. opinions (Electronic Word of Mouth, eWOM) on consumer attitudes and decision making when incorporated into an online movie streaming service. Third and last, we present the research design for a novel randomized controlled experiment (in progress) incorporating participants' Facebook data.

2   Theoretical Background and Related Work

2.1   Observational Learning

The theory of OL [2] has been applied in a number of different areas of information systems research, including analysis of IT adoption in the workplace [10, 13], IT skill acquisition [24], and decision-making in B2B markets [14]. OL can be described as learning which happens as a result of observing the behaviors of others. Basically, seeing someone perform an action increases the likelihood that an observer will also perform that action [9]. Because of the capability to learn by observation, humans do not always need to experience the outcome of actions themselves. Thus, OL relieves people of the tedious (and sometimes, hazardous) process of learning through trial and error [4].

OL is sometimes compared and contrasted with the impact of eWOM. For example, Chen et al. [7] compared the sales effect of positive and negative WOM with *"positive/negative OL information"*, and Libai et al. [16] called for more research into how WOM differ from OL. While we acknowledge the authors' attempts to separate opinions and behaviors, we argue that there is a need for further conceptual refinement of the mechanisms in place here. In the aforementioned examples, WOM and *"OL information"* are placed on the same level of analysis, namely as stimuli that potentially result in sales (or other outcome). We argue that that the concept of OL is not a stimulus but rather a process outcome. Consequently, we suggest that eWOM should be compared to eWOB, rather than to OL, and that OL is instead a potential outcome of eWOB.

2.2 Conceptual Framework for eWOB

Whether analyzed through the theoretical lens of OL or not, a diverse stream of research has begun investigating the increased observability of consumer behaviors and its effects. Figure 1 serves as an initial conceptual framework for eWOB (to be further developed in future research), including examples from extant research within this domain.

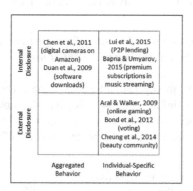

Fig. 1. Conceptual framework for eWOB and examples of extant research.

The first axis represents the level of aggregation in the behavioral information disclosed. Internet-based vendors such as Amazon often disclose information about the popularity of specific products, and which products are frequently bought together. Here, purchase behaviors of many individuals are aggregated when presented to other consumers. Contrary to this, we find services where information about specific individuals is disclosed. For example, information on Spotify about which songs one's connections have listened to. On the other axis, we distinguish between where the behavioral information is disclosed compared to where it originated. External disclosure happens when a product is consumed in one place, and a trace of that consumption behavior is disclosed on another platform. For example, listening to music in Spotify which generates a story shared to Facebook about this consumption behavior. Or when voting in an election (an offline "platform"), and claiming a "I just voted" badge on Facebook. Internal disclosure, on the other hand, happens when information about usage-behaviors is disclosed *within* the specific service, where usage took place. Back to the example of Spotify, internal disclosure is when information about the listening behaviors of a Spotify user is disclosed *within* Spotify, observable to other (connected) users of Spotify. In this paper, we will focus on how internal disclosure of individual-specific behavior-based information from Facebook friends can affect other users and potential users in terms of attitude towards a service and intention to use it.

3 Development of Hypotheses

Opinions as well as observable behavior of others are both examples of potential sources of social influence [12]. Thus, we first hypothesize that:

Hypothesis 1 (H1): *The inclusion of social information into a service leads to more positive attitude towards the service among potential users as well as higher intention to use the service.*

Where "social information" includes both opinion-based and behavior-based information from social others. Next, drawing on OL theory, we should also expect that social information about consumers' behaviors should have an impact, compared to seeing no social information:

H2: *The inclusion of behavior-based social information into a service leads to more positive attitude towards the service among potential users as well as higher intention to use the service.*

Building upon both OL theory [2] and Social Impact Theory [15], we recognize that a number of factors influence how effective the observational learning is, depending on: (1) who the model is (i.e. who performs the behavior) (2) how many persons perform that behavior, and (3) whether the outcome of the modeled behavior is observable or not.

Observable Outcome: Behavior vs. Opinions

Bandura [2] argues that the learning effect is amplified when the observant can observe the consequence(s) of the action. Thus, we expect that opinion-based social information will have a stronger impact than merely behavior-based, as opinions are assumed here to be formed on the basis of some prior action (behavior), and opinions thus reflect both a behavior *and* its outcome:

H3: *The inclusion of opinion-based social information into a service has a more positive impact on attitude towards service and intention to use service among potential users than including behavior-based social information.*

This is not to dismiss the potential of including behavior-based social information in products and services. While the above hypothesis may be proven true, the sheer amount of actions performed every day might altogether be more effective in influencing observers than the much smaller amount of actual reviews (observable outcome), as suggested in prior empirical research [1].

Type: Random vs. Category-specific People

Observing the actions of people who are knowledgeable about a specific topic, generally has a stronger impact than observing random/unknown people's behavior [2, 4, 15]. Consequently, we hypothesize that:

H4: *The inclusion of social information from product category-specific influential friends into a service has a more positive impact on attitude towards service and intention to use service among potential users than including social information from random friends.*

Number of People

Finally, we expect that observing a certain behavior among many friends will have greater impact on the individual than observing only one friend [15]. However, economic theory predicts that observing the actions of as few as two people can be the turning point where one starts disregarding own signals and starts following the behaviors of others [4]. Consequently, we hypothesize that:

H5: *Observing the behaviors and opinions of a low number of friends has a more positive impact on attitude towards service and intention to use service among potential users than observing no actions or opinions.*

Finally, because actions can reflect information to observers [4], observing a high number of people perform a behavior can lead to the inference among observers, that the behavior is in some way beneficial or attractive. Consequently, we hypothesize that:

H6: *Observing the behaviors and opinions of a high number of friends has a more positive impact on attitude towards service and intention to use service among potential users than observing the behaviors or opinions of a low number of friends.*

4 Research Design

In order to assess the impact of eWOB, we carried out a randomized controlled field experiment. The empirical context was an online movie streaming service in Denmark, more specifically Blockbuster[1] (BB). Movie streaming was chosen because of its socialness, and the online format which enabled integration with Facebook, and BB

---

[1] After the bankruptcy of the American movie rental chain Blockbuster, the rights to the Blockbuster brand in the Danish market were acquired by Danish telco TDC Group in 2014, and Blockbuster was re-launched as a movie streaming service.

proved to be interested in partaking in this research. The experiment was run as a $2 \times 2 \times 2$ factorial design with three independent variables each with two levels resulting in 8 treatment groups and a control group.

The experiment consisted of an online questionnaire and a website (stimuli). The experiment was run as post-test only where treatment groups were compared to a control group, which was exposed to a baseline version of the website without any integration of social elements.

4.1   Participants and Procedure

The target group for the experiment was potential new users (current non-users), of BB as we wanted to uncover how the inclusion of social information affected potential new users' attitude towards the service as well as their attention to start using it. Recruiting was done via promoted posts from BB Denmark's official Facebook page. We used the Facebook ad system combined with BB's own database to target users who matched BB's definition of potential new customers. Participants were randomly assigned upon entering the survey to one of eight treatment groups or the control group (Fig. 2).

Fig. 2. Experiment flow.

After clicking the Facebook post, participants were taken to an online survey. Participants were informed that the survey was run by Facebook and a group of researchers from a specific, well-known local university, and that the purpose was to get participants' opinion about a new version of the website. Participants (except control group) were also informed that the survey included a request to connect with Facebook. It was stressed that this was necessary to run the survey, and would only be used for this particular study, and that the permission would not allow us to post to Facebook on participants' behalf nor on the timelines of their friends. Once connected, half of these groups were shown a list of their Facebook friends and were asked to mark at least 10 friends who they thought had "an interesting movie taste". Selected friends were regarded as 'category-specific influentials'. It was stressed that the friends selected would not be notified about this selection. The other half of the treatment groups skipped this step. Next, all groups visited a mock-up version of the BB streaming website. The site was manipulated in 8 different combinations, cf. Table 1, utilizing Facebook profile photos of participants' friends to illustrate friends' (fictional) use of BB. Participants were instructed to click around on the site, and as a minimum look at three specific pages. As such, the task was not far from that facing a potential customer visiting the website for the first time. Upon that, participants returned to part 2 of the survey, which included the dependent variables plus a manipulation check for all treatment groups to uncover whether participants noticed the elements with social information. In addition to this, we used tracking software to track participants' mouse movements throughout their visit on the mockup website[2], as mouse movement can be used as a proxy for eye movement [6].

Table 1. Overview of independent variables and stimuli

Independent variable	Levels	Implication for stimuli
Type of social information	Behavior vs. opinions	Info on "Movies seen by your friends" vs. "Movies liked by your friends" and "These friends use Blockbuster" vs. "These friends like Blockbuster"
Type of friends	Random vs. category-specific influentials	Info from randomly picked friends vs. Pre-selected by participants as having "*an interesting movie taste*"
No. of friends	Low vs. high	2 friends vs. 8 friends shown

4.2 Dependent Variables

We measured attitude towards the movie streaming service using an adapted three-item approach based on existing measures of attitude towards a website [18, 23]. All items were measured on a 7-point Likert-style scale. As BB is a pay-per-view service, intention to start using the service (our second dependent variable) can be viewed as a purchase intention (PI). PI was measured using four items, all on a 7-point scale, and adapted from existing literature [3, 19].

5 Limitations and Future Research

At the time of submission responses from 473 people have been collected. The analysis hereof is still ongoing, and thus, the results are not reported here.

Like most research, this is not without limitations. On the conceptual side, we are currently working to refine the eWOB framework to include, amongst others, a time dimension (synchronous vs. asynchronous disclosure). In terms of our experiment, it was run on a mock-up of the real website of interest. This gave us opportunities to manipulate more variables but also carries the risk that participants will be shown friends who – to the participant - are very unlikely users of the streaming service investigated. This was not found to be an issue in any of the pilots, however it could blur the true effect of providing social information. In addition, since the friend information was fictional with regard to actual movie consumption behavior, it did not allow us to do network analysis to uncover diffusion among friends. Second, our method of identifying category-specific influential carries the risk that participants might feel obligated to answer consistently with this pre-selection. Or the contrary, they

---

[2] Participants were informed about the use of this software upon entering the questionnaire.

could also answer more negatively, because they don't want to signal that they could be under influence of their friends. Finally, other designs of the social integration into the website might lead to slightly different results and future research should test this. Related to this, future research should increase knowledge of what motivates consumers to disclose their product-related behaviors. This is an under-researched topic but nevertheless crucial in understanding the phenomenon of eWOB, and – from a managerial perspective – how products and services can be successfully designed for eWOB.

References

1. Aral, S., Walker, D.: Creating social contagion through viral product design: a randomized trial of peer influence in networks. Manage. Sci. **57**(9), 1623–1639 (2011)
2. Bandura, A.: Social Foundations of Thought and Action: A Social Cognitive Theory. Prentice-Hall, Englewood Cliffs (1986)
3. Bickart, B., Schindler, R.M.: Internet forums as influential sources of consumer information. J. Interact. Mark. **15**(3), 31–40 (2001)
4. Bikhchandani, S., et al.: Learning from the behavior of others: conformity, fads and informational cascades. J. Econ. Perspect. **12**(3), 151–170 (1998)
5. Bond, R.M., et al.: A 61-million-person experiment in social influence and political mobilization. Nature **489**(7415), 295–298 (2012)
6. Chen, M.C., et al.: What can a mouse cursor tell us more? In: CHI 2001 Extended Abstracts on Human Factors in Computing Systems - CHI 2001, p. 281. New York, ACM Press (2001)
7. Chen, Y., et al.: Online social interactions: a natural experiment on word of mouth versus observational learning. J. Mark. Res. **48**(2), 238–254 (2011)
8. Cheung, C.M.K., et al.: Do actions speak louder than voices? The signaling role of social information cues in influencing consumer purchase decisions. Decis. Support Syst. **65**, 50–58 (2014)
9. Cialdini, R.B.: Harnessing the science of persuasion. Harv. Bus. Rev. **79**(9), 72–81 (2001)
10. Compeau, D.R., et al.: From prediction to explanation: reconceptualizing and extending the perceived characteristics of innovating. J. Assoc. Inf. Syst. **8**(1), 409–439 (2007)
11. Duan, W., et al.: Informational cascades and software adoption on the internet: an empirical investigation. MIS Q. **33**(1), 23–48 (2009)
12. Fulk, J., et al.: A social influence model of technology use. In: Fulk, J., Steinfield, C. (eds.) Organizations and Communication Technology, pp. 117–141. SAGE Publications Inc., Thousand Oaks (1990)
13. Kim, C., et al.: An empirical investigation into the utilization-based information technology success model: integrating task- performance and social influence perspective. J. Inf. Technol. **22**(2), 152–160 (2007)
14. Koh, T.K., Fichman, M.: Multihoming users' preferences for two-sides exchange networks. MIS Q. **38**(4), 977–996 (2014)
15. Latané, B.: The psychology of social impact. Am. Psychol. **34**(4), 343–356 (1981)
16. Libai, B., et al.: Customer-to-customer interactions: broadening the scope of word of mouth research. J. Serv. Res. **13**(3), 267–282 (2010)
17. Liu, D., et al.: Friendships in online peer-to-peer lending: pipes, prisms, and relational herding. MIS Q. **39**(3), 729–742 (2015)

18. Lynch, P.D., et al.: The global internet shopper: evidence from shopping tasks in twelve countries. J. Advert. Res. **41**(3), 15–23 (2001)
19. Putrevu, S., Lord, K.R.: Comparative and noncomparative advertising: attitudinal effects under cognitive and affective involvement conditions. J. Advert. **23**(2), 77–90 (1994)
20. Rogers, E.M.: Diffusion of Innovations. Free Press, New York (2003)
21. Shi, Z., Whinston, A.B.: Network structure and observational learning: evidence from a location-based social network. J. Manage. Inf. Syst. **30**(2), 185–212 (2013)
22. Simpson, P.M., et al.: Understanding the consumer propensity to observe. Eur. J. Mark. **42** (1/2), 196–221 (2008)
23. Stevenson, J.S., et al.: Webpage background and viewer attitudes. J. Advert. Res. **40**(1–2), 29–34 (2000)
24. Yi, M.Y., Davis, F.D.: Developing and validating an observational learning model of computer software training and skill acquisition. Inf. Syst. Res. **14**(2), 146–169 (2003)

Locked in by Social Media Features?
Translating Clicks and Comments into Value

Eric C. Larson and Vidya Haran[✉]

University of Illinois at Urbana Champaign, Champaign, IL, USA
{ecl,vharan}@illinois.edu

Abstract. Social media platforms such as Facebook, LinkedIn, Twitter, and Snapchat have yielded huge valuations in equity markets and through acquisitions. In this study, we propose that the design features experienced by the platform users are important drivers of value in social media platforms. Also, we suggest a research design to empirically test the link between social media design feature types, the subjective value of the platform as assessed by users, and how platform features are reflected in equity value.

Keywords: Social media · Value · Design features

1 Introduction

Interest in the topic of social media is high, yet the attribution of value to the phenomenon is an elusive endeavor (Kaplan and Haenlein 2010). In a relatively short span of time, new types of social media platforms have come on the scene, many becoming instant sensations (e.g., WhatsApp, Snapchat, and Weibo), while many more became relative failures (e.g., MySpace, Google+, and del.icio.us). Academic research on this topic has primarily focused on the individual unit of analysis (e.g., Rishika et al. 2013; Hildebrand et al. 2013), with the outcome variable frequently related to individual adoption, user satisfaction (Hu et al. 2015), or individual purchase behavior (Chen et al. 2013; Chintagunta et al. 2010).

The impact of social media features extends beyond the individual and may create or erode value for the firm (Aral et al. 2013). Many firms miss opportunities to engage their customers through social media and fail to fully embrace what can be done to drive business results (Culnan et al. 2010). There are examples of how social media influences value as communicating using social media provides the ability to reach a wider audience (Chen et al. 2015) and research indicates that social media metrics are related to firm equity value (Luo et al. 2013).

It is important that social media **platforms** such as Facebook, LinkedIn, and Twitter are recognized as strategic actors and, thus, that an analysis of the impact of social media takes into account the impact on platforms (Aral et al. 2013). We address this call for research into this area by looking at drivers of value for these social media platforms. By doing so, we also address the concern that managers and executives face a relative dearth of information regarding strategic value creation at an organizational level (e.g. Goh et al. 2013). This research is designed to help fill this gap in the

© Springer International Publishing AG 2017
M. Fan et al. (Eds.): WeB 2016, LNBIP 296, pp. 104–109, 2017.
https://doi.org/10.1007/978-3-319-69644-7_10

literature by assessing specific design features of social media platforms from an individual level and understanding how those features translate into value for the user, and financial value for the social media platform at an firm level. The specific research question we ask is: How do social media design features affect the subjective user assessment of value, and how do these features reflect the objective market value of the platform?

2   Theoretical Background

In research focused on the interactions among users, social media networks are defined by the existence of a user profile, user access to digital content, transparent relations among users, and the ability to traverse those connections (Kane et al. 2014). Our definition parallels this notion of a social media network with respect to the user profile and transparent connections or relationships among those users, however our focus is on the social media platform (the technology and interaction of the user with the technology and other users). Hence, we emphasize the internet-enabled platform and the necessity of user-generated content (DesAutels 2011) in addition to the social interactions among users to define a social media platform. Additional design feature types (introduced below) enhance a user's experience with a social-media platform, but these features are not required for all platforms.

Design features are a fundamental building block in understanding the impact of social media on business (Aral et al. 2013). Various typologies exist in the literature regarding social media design features (Kietzman et al. 2011; Huang and Benyoucef 2013). Each design feature from these typologies type allows us to unpack and examine a specific facet of social media user experience through the platform's design features including identity, relationship content generation, formal groups, presence and reputation. Three of the design feature types (i.e., identity, relationships, and content generation) encompass features that help define whether a technology is indeed a social media platform. On the other hand, three feature types may exist, but are not required to define a technology as a social media platform.

3   Design Features and Value

The discussion of value is made difficult by pronounced differences in what value means to different people (Adler 1956). We use the extant literature on social media to operationalize the user assessment of value. Social media user value is a tradeoff between benefits (utilitarian and hedonic benefits) and cost (information risk, sacrifice, and effort) (Aral et al. 2013; Hu et al. 2015). Users evaluate utilitarian and hedonic benefits relative to what they sacrifice in effort and risk to derive a value determination of online social networking services (Hu et al. 2015).

In considering value, social media platform companies also must think deeply about design. The interfaces, policies, and features that the platform provides not only structures how users interact, but also how third parties can provide add-on features and applications that extend a platform's functionality (Aral et al. 2013). Specifically with social media platforms, value often comes from the right balance of features and subsidizing the product early to encourage adoption (Dou et al. 2013). "Features of the particular platform cause certain types of social interactions to flourish, more so than others" (Kane et al. 2014: 279) and as a result can drive value for the platform. Users also consider the tradeoff between privacy and personalization and will share more personal information when features are present to help the user know what is being done with the information once it is provided (Awad and Krishnan 2006). And social media platforms must also recognize that users devise new ways of using software features that extend beyond the original intent (Sun 2012).

Our propositions about design features, and the value of social media platform are:

Proposition 1: A user's subjective assessment of value increases as design features move along a spectrum based on whether the feature enables functionality for the user only, dyadic interaction or interaction among a group.
Proposition 2: User's social judgment of value will understate the increasing trend from individual to dyad to network levels.

The second proposition remains an empirical question, but we expect that user's will underestimate the increase in value of design features as they move up the levels. Our proposition predicts that users will perceive design features as approximately equal in value and fail to fully recognize the value to the social media platform of sharing user-generated content (DesAutels 2011).

4   Methodology

Data for this study will be collected in a survey questionnaire given to undergraduate and graduate students at a major U.S. university. While it is true that social media users come from all age categories, individuals between the ages of 18 and 29 have the highest adoption rates of any age group on Facebook (84% in 2013 and 87% in 2014 respectively) and Twitter (31% and 37%), while that age category is the fasted growing segment of users on LinkedIn (Duggan et al. 2015). College aged youth 18–24 make up nearly half of Snapchat users (Hoelzel 2015). We will supplement the survey data with publicly available financial data from Compustat and company annual reports to provide the equity valuation for each social media platform.

Features of four publicly traded social media platforms (Facebook, Twitter, LinkedIn, and Snapchat) will be included in the study. These platforms (except Snapchat) were selected because they are among the most popular social media plat-forms or, with regard to Snapchat, it is the fastest-growing social media app. The platform has grown from 30 million active users in 2014 (Oremus 2015) to more than 150 million active daily users in 2016 (Frier 2016). Also, Snapchat has a unique design feature of ephemeral communication with no permanent record of activity, making it appealing to youngsters who fear the permanency of social media platforms such as Facebook (Canal 2014). The identification of features on the platform will be done systematically to be as comprehensive as possible. It will include a complete review of the social media system menus and support/help menus. The identification process will also capture screen shots of each design feature to provide rich visual cues for respondents during the survey process.

The overall research design will include the following steps: (1) Identify the design features of four major social media platforms (Facebook, Twitter, LinkedIn, and Snapchat); (2) Code the design features by q-sorting performed by multiple independent analysts to categorize each design feature into the appropriate design feature type from our typology of social media features; (3) Run a test devised to assess the clarity of the questions, to ensure that the questionnaire will provide appropriate measurement validity, and to evaluate the most used features for inclusion in the large-scale survey; (4) Survey 300 social media users regarding their assessment of social media design features and subjective assessment of social media platform value; and (5) Match the survey data to financial data from Compustat, annual reports, and other publicly available sources to provide equity valuation figures and other control variables.

5 Discussion and Conclusion

Our first contribution is to define social media platforms according to the underlying technology and the design features that the user experiences. This definition helps to form the boundaries and uncover opportunities for research in the nascent area of social media. From this definition, we create a new typology of design feature types which exist on a variety of social media platforms. This typology not only allows the classification of social media design features by feature type, but it also provides a foundation and insight into why and how design features differ in value for the social media platform. Finally, we will empirically investigate the relationship between specific social media design features and financial value.

It is no doubt helpful for platforms such as Facebook, Twitter, LinkedIn and Snapchat to understand what features and feature types create the most value for the platform. This critical information could influence managers in the adoption of specific design features and may guide social media platform executives in developing future versions of the platform. Some specific design features may prove to be substantial firm assets that should be considered strategically by top executives of the social media platform.

By comparing design features across four popular platforms, we expect to be able to answer questions on whether these platforms serve as supplements or complements to each other in satisfying user needs, thus responding to the call for research in this area as outlined by Aral et al. (2013).

At the same time, users benefit from a healthy appreciation of the relative value they create for the platform through their use of specific design features. This appreciation ensures that users have a strong position vis-à-vis the relationship they have with social media platforms so that users make the most appropriate usage choices. Greater transparency about the financial implications of feature usage also helps users make informed decisions about their own social media usage.

References

Adler, F.: The value concept in sociology. Am. J. Sociol. **62**(3), 272–279 (1956)

Aral, S., Walker, D.: Creating social contagion through viral product design: a randomized trial of peer influence in networks. Manag. Sci. **57**(9), 1623–1639 (2011)

Aral, S., Dellarocas, C., Godes, D.: Introduction to the special issue—social media and business transformation: a framework for research. Inf. Syst. Res. **24**(1), 3–13 (2013)

Awad, N.F., Krishnan, M.S.: The personalization privacy paradox: an empirical evaluation of information transparency and the willingness to be profiled online for personalization. MISQ **30**(1), 13–28 (2006)

Canal, E.: The inside story of snapchat: the world's hottest app or a \$3 billion disappearing act. forbes.com (2014). http://www.forbes.com

Chen, Y., Qang, Q., Xie, J.: Online social interactions: a natural experiment on word of mouth versus observational learning. J. Mark. Res. **48**(2), 238–254 (2013)

Chen, H., De, P., Hu, Y.J.: IT-enabled broadcasting in social media: an empirical study of artists' activities and music sales. Inf. Syst. Res. **26**(3), 513–531 (2015)

Chintagunta, P., Gopinath, S., Venkataraman, S.: The effects of online user reviews on movie box office performance: accounting for sequential rollout and aggregation across local marketing. Mark. Sci. **29**(5), 944–957 (2010)

Culnan, M.J., McHugh, P.J., Zubillaga, J.I.: How large US companies can use Twitter and other social media to gain business value. MISQ Executive **9**(4), 243–259 (2010)

DesAutels, P.: Understanding the nature of user-generated information systems. Bus. Horiz. **54**, 185–192 (2011)

Dou, Y., Niculescu, M.F., Wu, D.: Engineering optimal network effects via social media features and seeding in markets for digital goods and services. Inf. Syst. Res. **24**(1), 164–185 (2013)

Duggan, M., Ellison, N.B., Lampe, C., Lenhart, A., Madden, M.: Demographics of key social networking platforms (2015). http://www.pewinternet.org/2015/01/09/demographics-of-key-social-networking-platforms-2/. Accessed 5 May 2016

Frier, S.: Snapchat Passes Twitter in Daily Usage. Bloomberg Technology (2016). http://www.bloomberg.com/news/articles/2016-06-02/snapchat-passes-twitter-in-daily-usage. Accessed 2 June 2016

Goh, K.Y., Heng, C.S., Lin, Z.: Social media brand community and consumer behavior: quantifying the relative impact of user- and marketer-generated content. Inf. Syst. Res. **24**(1), 88–107 (2013)

Hu, T., Kettinger, W.J., Poston, R.S.: The effect of online social value on satisfaction and continued use of social media. Eur. J. Inf. Syst. **24**(4), 391–410 (2015)

Huang, Z., Benyoucef, M.: From e-commerce to social commerce: a close look at design features. Electron. Commer. Res. Appl. **12**(4), 246–259 (2013)

Hildebrand, C., Häubl, G., Herrmann, A., Landwehr, J.R.: When social media can be bad for you: community feedback stifles consumer creativity and reduces satisfaction with self-designed products. Inf. Syst. Res. **24**(1), 14–29 (2013)

Hoelzel, M.: A breakdown of the demographics for each of the different social networks. Business Insider (2015). http://www.businessinsider.com/update-a-breakdown-of-the-demographics-for-each-of-the-different-social-networks-2015-6. Accessed 5 May 2016

Kane, G.C., Alavi, M., Labianca, G., Borgatti, S.: What's different about social media networks? A framework and research agenda. MISQ **38**(1), 275–304 (2014)

Kaplan, A.M., Haenlein, M.: Users of the world, unite! The challenges and opportunities of social media. Bus. Horiz. **53**(1), 59–68 (2010)

Kietzmann, J.H., Hermkens, K., McCarthy, I.P., Silvestre, B.S.: Social media? Get serious! Understanding the functional building blocks of social media. Bus. Horiz. **54**(3), 241–251 (2011)

Luo, X., Zhang, J., Duan, W.: Social media and firm equity value. Inf. Syst. Res. **24**(1), 146–163 (2013)

Rishika, R., Kumar, A., Janakiraman, R., Bezawada, R.: The effect of customers' social media participation on customer visit frequency and profitability: an empirical investigation. Inf. Syst. Res. **24**(1), 108–127 (2013)

Oremus, W.: Is Snapchat Really Confusing, or Am I Just Old? (2015). Slate.com Available at http://www.slate.com/articles/technology/technology/2015/01/snapchat_why_teens_favorite_app_makes_the_facebook_generation_feel_old.html

Sun, H.: Understanding user revisions when using information system features: adaptive system use and triggers. MISQ **36**(2), 453–478 (2012)

When Your App is Under the Spotlight

Chen Liang[(✉)], Zhan (Michael) Shi, and T.S. Raghu

Arizona State University, Tempe, AZ 85287, USA
chen.liang.4@asu.edu

Abstract. Due to the nature of apps as experience goods and the vast product space, new product discovery in the mobile app market has become a very salient problem for developers and consumers. Our paper investigates developers' best response to the platform's recommendation, where the featured apps enjoy a reduction in search cost and an endorsement of quality for a limited period of time. Specifically, we consider three response options available to the developers: releasing a version update, increasing price, and decreasing price. We find that only the price decrease strategy has a positive effect on sales during the featuring window, while the effect of the price increase and version update strategy is not significant.

Keywords: Platform recommendation · Signaling strategy · Search cost · Product discovery · Mobile app market

1 Introduction

App distribution platforms such as the Apple iTunes App Store and Google Play Store that connect developers and consumers of mobile apps have seen tremendous growth in recent years. According to industry statistics, around 2 million apps have been released in the two stores, and it is projected that the total size of the "app economy" may reach 101 billion dollars by 2020.[1] Given the rapidly expanding market size and the declining cost of entry, more and more app developers join this market, which reinforces the growth in the demand. However, though consumers have installed an estimated 156 billion mobile apps on their devices in 2015,[2] the distribution of these downloads is highly skewed – over 90% of the apps are downloaded less than 500 times per day.[3] Because of such a concentrated market structure and the potential information overload problems in mobile commerce [1], new product discovery has become a very salient problem for developers and consumers.

On the app distribution platforms, the primary organic channel of new app discovery is search. According to the estimation from Nielsen [2], direct search is the prevalent method for discovering new apps. To optimize search, the developers need to improve the quality of their apps, strategically choose the search terms to compete for,

---

[1] See http://www.statista.com/statistics/276623/number-of-apps-available-in-leading-app-stores/, and http://venturebeat.com/2016/02/10/the-app-economy-could-double-to-101b-by-2020-research-firm-says/.

[2] See https://www.idc.com/getdoc.jsp?containerId=prUS41240816.

[3] See http://www.gartner.com/newsroom/id/2648515.

© Springer International Publishing AG 2017
M. Fan et al. (Eds.): WeB 2016, LNBIP 296, pp. 110–116, 2017.
https://doi.org/10.1007/978-3-319-69644-7_11

and carefully craft their app title, description, and other listing information so as to obtain a high ranking position in the search result page. Besides the metadata in the store listing, the platform also uses existing downloads and user feedback in determining the search ranking. New apps, particularly those released by independent developers, usually lack the resource needed for conducting a marketing effort to attract a substantial initial user base. Therefore, even high quality new apps often find it extremely difficult to achieve substantive download growth in the search channel.

A key emerging channel of new product discovery is the platform's recommendation. For instance, Apple's App Store editors select and recommend new and recently updated apps in the featured apps column of its home page every day. When featured by the store, these new apps get more exposure to potential consumers, which suggests an exogenous reduction in search cost. Meanwhile, the store's recommendation also serves as an endorsement, which increases consumers' expectation of the apps' quality. Considering the reduction in search cost and endorsement of quality under the spotlight, what is the developers' best response strategy? Given the limited action set, should developers release a version update to signal developer support? Should developers increase app price to enforce the high quality signal? Or should they provide a temporary price discount in order to boost downloads?

While a large body of research has focused on the effectiveness of recommendation [3–6] and the design of recommendation systems [7–9], few studies have considered the developers' best response to the exogenous recommendations. The primary objective of this study is to answer the following question:

When their products receive an exogenous reduction in the search cost and a quality endorsement from the platform's recommendations, what is the developers' best response strategy?

2 Theoretical Discussion

Since consumers cannot ascertain the value of an app before actually using it, they form an expectation of the app quality according to available information. In order to overcome such information asymmetry issues, sellers resort to signaling mechanisms to differentiate themselves from other competitors [10]. Previous literature finds that price [10], recommendations [5], reviews [11], brands [12] are all effective product signals. Specifically, two broad categories of product signals exist in the app distribution platforms, namely, platform-generated signals and seller-generated signals. Platform-generated signals include platform recommendations, search result rankings, and any sales or download rankings. The literature has shown that the platform's recommendations are regarded as strong quality signals which distinguish the high-quality products from the low-quality products [5, 13], while powerful, platform-generated signals are by definition out of the sellers' direct control. On the other hand, seller-generated signals are product cues provided by sellers to help users better evaluate the product quality and the fit between the product and users' preference. In the mobile app markets, price increase, price decrease, and version update are all prime examples of seller-generated signals. First, since high price is often perceived to be associated with high quality, developers may choose to increase app price to signal that their apps are superior to the competitors. Second, by providing a relatively low introductory price, decreasing price serves as a product promotion signal [10]. By sending such product promotion signals, sellers communicate to potential users that they have confidence in their capability to attract a large user base in order to recover the short-term loss (the promotional expense) [10]. Third, releasing a version update signals that developers have improved the functionality of the apps based on the previous reviews and market surveys or they have included new features in the updated version [14]. Therefore, a version update is also a salient quality signal to help sellers to distinguish from other competitors. Overall, the platform-generated signals and seller-generated signals are mainly employed to convey to users the product's quality and promotion strategy.

During the time window when the platform recommendation is sending a strong quality signal on the featured apps, the developers' best response strategy should be determined according to the complementarity of platform-generated signal and seller-generated signal. It's reported by the prior literature that multiple quality signals might substitute for each other, and thus, the dominating strategy is to send either signal [15]. Therefore, we expect that when the strong platform-generated quality signal is already provided, seller-generated quality signals might not have a significant effect on demand. Specifically, when the apps are recommended by the platform, neither increasing the price nor releasing a version update will have a positive effect on the sales or download performance. However, since the product promotion signals highlight on the sellers' long-term plan and confidence, they might complement with quality signals which focus on the product quality. Hence, when the apps are recommended by the platform, decreasing the price might augment the effect of the platform's recommendations. Based on the relationship between multiple signals, we propose the following hypotheses:

H1: Given the platform's recommendation, increasing the app price will not improve the download performance.
H2: Given the platform's recommendation, releasing a version update will not improve the download performance.
H3: Given the platform's recommendation, decreasing the app price will improve the download performance.

3   Research Methodology

3.1   Data

In order to answer this research question, we collect our data from Apple's iTunes App Store (the Store hereafter). The dataset consists of three parts. The first part is the focal apps of our study that were featured in the Store's "New Apps We Love" column[4] between February 1, 2016, and July 31, 2016. Since iTunes does not provide an API endpoint for retrieving the list of featured apps programmatically, we first took

---

[4] The section used to be named "Best New Apps.".

screenshots of the Store's homepage and the complete feature section every day and manually identified the featured apps from the images. We then periodically checked each identified app's historical data on the mobile app market intelligence website App Annie[5] to make sure the manually identified app list matches with the archived information provided there.[6] The featuring window varies from 6 days to 21 days with a mean equal to 9.7 days. We provide descriptive statistics of selected app attributes in Table 1.

Table 1. Summary statistics of some selected attributes of the featured apps

Variable	Variable definition	Mean	Std. Dev.	Min	Max
Free	A dummy variable, = 1 if the app is free	0.72	0.45	0	1
Price	The average app price during our observational period	5.62	16.62	0	149.99
Avg rating	The average rating for the app	4.03	0.64	1.5	5
Rating count	Number of ratings for the app	1187.35	8914.98	5	148303
Tenure	The tenure measured in days when the app was featured for the first time	205.68	419.04	0	2565

The dataset's second component comes from the store listing information provided by the iTunes API. For the featured apps, we collected both their static listing attributes, such as title, developer, category and recommended age group, and dynamic information including version and price (for paid apps). We continuously monitored these app change events since January 1, 2016. The third part of the dataset is the Store's "Top Free" and "Top Paid" charts, which we also tracked daily since January 1, 2016. For each day, we observe the 100 top-ranked free apps and 100 top-ranked paid apps in the whole Store as well as in each of the 24 categories designated by iTunes. Note that the Store has country-specific versions and our programs and manual data collection tasks were all implemented in the U.S., so our data of featured apps and top charts are specific to the U.S. Store (Table 2).

As a research-in-progress, we are in the process of collecting data on the featured apps' social network activity, on Twitter and Facebook, which will provide us a more complete view on the developers' promotion activity. We have also identified the Android version released in the Google Play Store for a subset of the featured apps. We are still in the process of collecting the Android apps' attributes, dynamic change events, and performance data. The purpose of collecting data on the same apps in the second app store is to test any spillover effect across platforms.

---

[5] See https://www.appannie.com/tours/market-data-intelligence/.

[6] For example, the link https://www.appannie.com/apps/ios/app/anchor-lets-talk/features/#device= iphone shows the information on store featuring for the app named "Anchor - Radio by the people.".

Table 2. Definitions and summary statistics of the key variables

Variable	Variable definition	Mean	Std. Dev.	Min	Max
toprank_dummy$_{it}$	A dummy variable(0, 1), = 1 if the app is listed in the top chart on day t	12.08	16.20	1.00	95.00
feature_dummy$_{it}$	A dummy variable(0, 1), = 1 if the app is featured on day t	9.39	4.88	1.00	21.00
increase$_{it}$	A dummy variable(0, 1), = 1 if the app increases its price on day t	0.02	0.14	0.00	1.00
decrease$_{it}$	A dummy variable(0, 1), = 1 if the app decreases its price on day t	0.02	0.13	0.00	1.00
update$_{it}$	A dummy variable(0, 1), = 1 if the app releases an update on day t	0.04	0.19	0.00	1.00

3.2 Analysis and Results

In order to investigate our hypotheses, we specify the following model for paid apps:

$$
\begin{aligned}
toprank\_dummy_{it} = {}& \alpha_0 toprank\_dummy_{it-1} + \alpha_1 feature\_dummy_{it} \\
& + \alpha_2 decrease_{it} + \alpha_3 decrease_{it} * feature\_dummy_{it} \\
& + \alpha_4 increase_{it} + \alpha_5 increase_{it} * feature\_dummy_{it} \\
& + \alpha_6 updated_{it} + \alpha_7 updated_{it} * feature\_dummy_{it} \\
& + \alpha_8 decrease_{it-1} + \alpha_9 increase_{it-1} + \alpha_9 update_{it-1} \\
& + \alpha_{10} app_i + \varepsilon_{it}
\end{aligned}
\tag{1}
$$

Since the price change strategies are only available to paid apps, we specify the following model for the featured free apps:

$$
\begin{aligned}
toprank\_dummy_{it} = {}& \alpha_0 toprank\_dummy_{it-1} + \alpha_1 feature\_dummy_{it} \\
& + \alpha_2 update_{it} + \alpha_3 updated_{it} * feature\_dummy_{it} \\
& + \alpha_4 updated_{it-1} + \alpha_5 app_i + \varepsilon_{it}
\end{aligned}
\tag{2}
$$

where *app$_i$* denotes the app-level fixed effect and ε_{it} denotes the error term. Taking the apps which didn't deploy any strategical responses as the baseline (without price changes nor version updates), we expect that α_1 is significantly positive, which implies the positive effect of platform's recommendations. More importantly, our tests for the hypotheses hinge on the interaction term between three optional strategies and the feature dummy. If the interaction term is significantly positive, it means that such a strategical response has a positive effect on the download performance and profit (Table 3).

Based on the result of the linear Fixed Effects model, we find that for both paid and free apps, the platform's recommendations have a substantial effect on the download performance. Moreover, the version update strategy also has a positive main effect on

Table 3. Estimation results of the fixed-effects model

DV : $toprank\_dummy_{it}$	Paid apps	Free apps
$toprank\_dummy_{it-1}$	0.701*** (0.025)	0.715*** (0.016)
$decrease_{it}$	−0.078* (0.041)	
$feature\_dummy_{it}$	0.219*** (0.021)	0.186*** (0.015)
$decrease_{it} \times feature\_dummy_{it}$	0.148* (0.079)	
$increase_{it}$	−0.016 (0.037)	
$increase_{it} \times feature\_dummy_{it}$	0.053 (0.077)	
$update_{it}$	−0.000 (0.028)	0.041*** (0.010)
$update_{it} \times feature\_dummy_{it}$	0.055 (0.051)	−0.008 (0.031)
$decrease_{it-1}$	0.017 (0.059)	
$increase_{it-1}$	0.022 (0.043)	
$update_{it-1}$	0.071*** (0.022)	0.054*** (0.010)
Intercept	0.079*** (0.006)	0.032*** (0.002)
Number of observations	5,730	14,048
Number of groups	84	211
Adj R-squared	0.642	0.712

Notes: a. Robust standard errors are reported in parentheses; b. *
$p < 0.1$, ** $p < 0.05$, *** $p < 0.01$.

the download performance. Specifically, we find that there is some time delay in the positive effect of version update on the paid apps' download performance. But for free apps, the version update can immediately improve the present and future down performance. However, the interaction term between the price decrease and the feature dummy is not significant, which lends support to Hypothesis 2. Regarding the pricing strategies for paid apps, the main effect of a price reduction is negative. Therefore, beyond the featuring window, lowering the app price has a negative effect on the download performance. However, the interaction term between the price decrease and the feature dummy is significantly positive. Taking both the main effect and the interaction term into consideration, a reduction in price leads to an increase in the download performance (0.148–0.078 = 0.070). Thus, Hypothesis 3 is supported. However, neither the main effect nor the interaction term of price increase is significant. So Hypothesis 1 is also supported by our data. Overall, our three hypotheses are all supported.

To sum up, we find that when the platform recommendation is sending a strong quality signal on the featured apps, neither the price increase strategy nor the version update strategy has a significant positive effect on the download performance, which implies the substitution relationship between multiple quality signals. However, decreasing the app price significantly boosts downloads, which indicates the complementary relationship between product promotion signals and quality signals. Therefore, during the time window when the platform recommendation is provided, the developers' best response strategy should be decreasing the app price.

References

1. Ghose, A., Han, S.P.: Estimating demand for mobile applications in the new economy. Manag. Sci. **60**(6), 1470–1488 (2014)
2. The Nielsen Company: Nielsen mobile apps white paper, September 2010
3. Chen, P.-Y., Wu, S., Yoon, J.: The impact of online recommendations and consumer feedback on sales. ICIS 2004 Proc. p. 58 (2004)
4. Lin, Z., Goh, K.-Y., Heng, C.-S.: The demand effects of product recommendation networks: an empirical analysis of network diversity and stability
5. Oestreicher-Singer, G., Sundararajan, A.: Recommendation networks and the long tail of electronic commerce. MIS Q. **36**(1), 65–83 (2010)
6. Pathak, B., Garfinkel, R., Gopal, R.D., Venkatesan, R., Yin, F.: Empirical analysis of the impact of recommender systems on sales. J. Manag. Inf. Syst. **27**(2), 159–188 (2010)
7. Adomavicius, G., Tuzhilin, A.: Toward the next generation of recommender systems: A survey of the state-of-the-art and possible extensions. IEEE Trans. Knowl. Data Eng. **17**(6), 734–749 (2005)
8. Komiak, S.Y., Benbasat, I.: The effects of personalization and familiarity on trust and adoption of recommendation agents. MIS Q. **30**(4), 941–960 (2006)
9. Xiao, B., Benbasat, I.: E-commerce product recommendation agents: use, characteristics, and impact. MIS Q. **31**(1), 137–209 (2007)
10. Kirmani, A., Rao, A.R.: No pain, no gain: a critical review of the literature on signaling unobservable product quality. J. Mark. **64**(2), 66–79 (2000)
11. Zhu, F., Zhang, X.: Impact of online consumer reviews on sales: the moderating role of product and consumer characteristics. J. Mark. **74**(2), 133–148 (2010)
12. Rao, A.R., Qu, L., Ruekert, R.W.: Signaling unobservable product quality through a brand ally. J. Mark. Res. **36**, 258–268 (1999)
13. Smith, D., Menon, S., Sivakumar, K.: Online peer and editorial recommendations, trust, and choice in virtual markets. J. Interact. Mark. **19**(3), 15–37 (2005)
14. Lee, G., Raghu, T.S., Park, S.: Do app descriptions matter? evidence from mobile app product descriptions. SSRN Working Paper (2015)
15. Milgrom, P., Roberts, J.: Price and advertising signals of product quality. J. Polit. Econ. **94**(4), 796–821 (1986)

A Social Endorsing Mechanism for Mobile Coupons

Yung-Ming Li[1], Jyh-Hwa Liou[1,2(✉)], and Ching-Yuan Ni[1]

[1] Institute of Information Management, National Chiao Tung University,
1001 University Road, Hsinchu 300, Taiwan, Republic of China
yml@mail.nctu.edu.tw, alioujh@gmail.com,
haha54carol@gmail.com
[2] Center for General Education, Hsin Sheng College of Medical Care
and Management, Taoyuan, Taiwan

Abstract. With the popularity of social media and mobile device, businesses have more digital coupon dissemination channels than before. However, coupon redemption rates are unsatisfactory to businesses. The objective of this study is to design a new social coupon endorsing mechanism, which aims to find out those endorsers who have coupon proneness and high sharing willingness in coupon diffusion, with a recommended list of coupon receivers. Our mechanism propagates mobile coupon or discount information in an efficient way by helping consumers effortlessly gather the coupons fitting their preference and location.

Keywords: Mobile coupon · Social network · Location-based commerce · Social endorser · Social diffusion

1 Introduction

Since social networking sites and mobile media have played a significant role, businesses have more coupon dissemination channels than before. A new report from Juniper Research, the number of mobile user will reach 1.05 billion in 2019 [1]. 49% of US smartphone owners have used mobile coupons on their devices [2]. In 2009, the amount of mobile coupons has skyrocketed and jumped 263% [3]. Blooming mobile coupons applications, such as Cellfire, Grocery IQ, Coupon.com and Saving star, are coming up, which provide abundant variety of coupons for consumers [4]. Mobile coupons offer huge business opportunities for marketer. Because of the growing coupon distribution channel, location-based coupon offers another chance to reach prospect consumers instantly.

As social media and mobile device moves to the center of the electronic commerce, they become an important medium to acquire prospective consumers, to influence shopper behavior through their social network, and to provide instant location-based service. In the increasingly competitive circumstances, currently mobile coupon providers are facing numerous challenge and difficulty. There are some major research problems still unsolved:

© Springer International Publishing AG 2017
M. Fan et al. (Eds.): WeB 2016, LNBIP 296, pp. 117–124, 2017.
https://doi.org/10.1007/978-3-319-69644-7_12

- How to help target consumers in a convenient and time-saving way to gather the coupons they need and the providing store is nearby.
- How to help marketers to disseminate mobile coupon through social networking sites without evoking consumers' irritation?

The objective of this study is to design a new social coupon endorsing mechanism, which aims to find out those endorsers who have coupon proneness and sharing willingness in coupon diffusion. Specifically to say, it will recognize those people who are easily motivated by rewards, sensitive on discount information and willing to sharing. Consequently, endorsers are more willing to actively diffuse coupons when they received coupons with reward. And with a given list of coupon receivers recommended by our mechanism, the endorsers are able to know to whom they can diffuse a coupon.

This paper is organized as follows. The basic concepts and literature related to our research topics are provided in Sect. 2. In Sect. 3, we present the system architecture of social couponing diffusion mechanism. Section 4 describes the processes of the experiment and discusses the empirical results. Section 5 provides several concluding remarks and future research.

2 Related Work

2.1 Mobile Couponing

Mobile coupon is a type of digital coupon that are commonly used for sales promotion issued by retailers and marketers, which aim to retain old customers while attracting new customers to lead them to take account alternative products and enjoy their benefits at checkout [5, 6]. Mobile coupon can be found from online websites, email, SMS and mobile applications. Consumers can easily carry with saving into their mobile devices and receive incentive at the time of redemption without requiring taking time to clip paper coupon. New research unveiled from GfK in 2013 shows that brand marketing using digital coupon can increase customer recall and purchase intention and it also lures digital coupon users to come to the store purchasing more frequently and spend 42% more than average customers [7]. Mobile devices have played a role of bridging the instant service gap between consumers and local retailers [8]. Mobile phone enhances the capabilities of advertisement reachability at right time and right place. Banerjee and Dholakia [9] addresses that the characteristics of location based mobile advertisements are significant factors to influence the effectiveness of an ad.

2.2 Social Diffusion

According to a report by the Pew Research Center, about 65% of American who gets news from social media sites and Facebook is the majority [10]. For this reason, widely delivering the right information to the right receiver is a significant issue for marketers.

The way of information diffusion plays an important role in affecting consumers' preference.

In the social network, our interpersonal relationship can be visible to public, and can be used to simplify the path of information spreading among individuals and study the behavior of social influences [11]. Social influence refers to person's perception, attitude, or behavior are changed by others [12, 13]. It has always been a topic of social psychology, which studies to explain how individuals are affected and become unfiled into a group [14, 15]. Many local retailers adopt SNS as a tool to advertise their shops. They encourage consumers to "like" brand or "check-in" to the business on Facebook, by providing reward incentives or promotion [16]. These actions are equivalent to give a good evaluation (recommendation) to the business from consumer's SNSs. In another aspect, consumer's friends will see the information on Facebook. If they are also interested in the store or products, that may increase the impact on the recommendation and stimulate their purchasing desire; meanwhile, business do not need to advertise themselves while product and discount information pass through by consumers' social network in a low cost but more efficiency way [17].

3 The Model

In this research, we will propose a social diffusing mechanism to disseminate effectively mobile coupon via social endorsers to support marketing strategies. When a consumer comes nearby the store or live in the same vicinity, they can receive an appropriate mobile coupon recommended from the friends with powerful influence on him/her. Besides, when the target consumers receive a location-based coupon in line with their need, it will stimulate them to redeem them and forward the coupon to their friends for acquiring rewards. In the process of coupon diffusion, we expect the target consumers will be influenced by current endorsers, and eventually become the next coupon endorser. The system architecture is shown as Fig. 1. The components included in the systems are described as follows.

3.1 Target Consumer Discovering Mechanism

Preference Analysis Module. Brand is an important factor to drive coupon receivers to redeem coupon. A consumer who is with brand loyalty is a good target to promote the coupon. We can easily identify the user's brand preference from his/her Facebook feature "Like". In social networking site, such as Facebook, users who have pressed "Like" on the fan page will be stored. Once the fan pages have something news updated, they will also be shown in fans' Facebook wall such that business can stay connection with their fans.

Coupon Attribute Analysis Module. We adopt a superior distance-based method [15] to calculate the similarity between the keywords of consumers' preference and coupon attribute. Note that $\alpha \geq 0$ is a constant and $\beta > 0$ is a smoothing factor, and referenced to [18] we set $\alpha = 0.2$ and $\beta = 0.6$. The similarity score is formulated as below. The category tree is shown as Fig. 2.

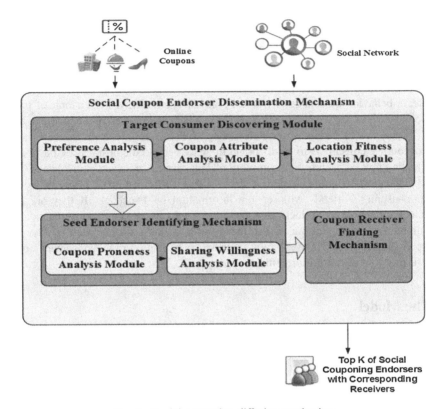

Fig. 1. Social couponing diffusion mechanism

$$Sim(C_1, C_2) = e^{-\alpha l} \times \frac{e^{\beta h} - e^{-\beta h}}{e^{\beta h} + e^{-\beta h}} \qquad (1)$$

Location Fitness Analysis Module. The preference is a significant factor when a consumer considers coupon redemption. However, if the redemption service location is far away from them, the coupon still cannot drive them to use. This module measures the similarity degree between a user's location and the coupon redemption location providing a relevant coupon service.

When mobile coupons are recommended to consumers, location is one of an important factor to drive consumer to redeem coupons. We consider three parts of redemption factors which can effect consumers' willingness: current location $C_{u_i,c}$. As a result, the feasible location score can be calculated by the following formula:

$$L(u_i, c) = C_{u_i,c} \qquad (2)$$

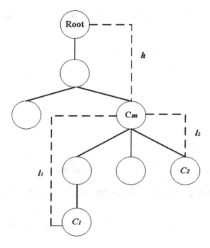

Fig. 2. Category tree

By considering consumers' preference and available location, we can further compute the fitness score by using the following formula:

$$Fitness(u_i, c) = Sim_p(u_i, c) + L(u_i, c) \tag{3}$$

3.2 Seed Endorser Identifying Mechanism

Coupon Proneness Analysis Module. The static proneness of a target user t_i can be calculated by following formula:

$$SP(t_i) = gender_{t_i} \times \left(\sum_{j=0}^{n} count(t_i, p_j) \right) \tag{4}$$

$$Coupon\_Prone(t_i) = SP(t_i) \tag{5}$$

3.3 Coupon Receiver Identifying Mechanism

The interactions on social media include response, like, tag, and share message posted by other user. Given coupon endorser e_i and target consumer t_j, the social interaction score is formulated as:

$$SI(e_i, t_j) = \frac{|\Phi interaction(e_i, t_j)|}{|\Phi interaction(e_i)|} \tag{6}$$

Note that $\Phi interaction(e_i)$ indicates the total number of social activities e_i exhibiting on social media. And the $\Phi interaction(e_i, t_j)$ indicates that total number of social activities with both e_i and t_j. The greater the social interaction score, the more the closeness between two users.

Finally, the seed endorser and the suitable coupon receiver can be matched by prior equations:

$$Match(e_i, t_j) = SI(e_i, t_j) \tag{7}$$

4   Experiment

4.1   Experimental Design

The research experiment procedures are described as follows.

(1) Identifying the target consumers from the data collected from participants' social network.
(2) Selecting the coupon endorsers for disseminating mobile coupons by different planned strategies.
(3) The diffusion reaction is recorded into database. Note that when endorsers receive a coupon, they do not know which coupon dissemination approach to be tested.
(4) After endorsers visit the given webpage and make a decision on whether to pass coupon, the endorsers have to respond their reactions on this coupon by online questionnaire.

4.2   Experimental Results

Average Sharing Times. The average sharing times is a significant factor to quantify the effectiveness of the coupon diffusion process. Once the coupon delivered from endorser to target consumer, we can measure the times of a coupon was shared to next target consumers. The coupon average sharing times (AST) formula is defined as (Fig. 3):

$$AST = \frac{\Phi shared}{\Phi coupon}$$

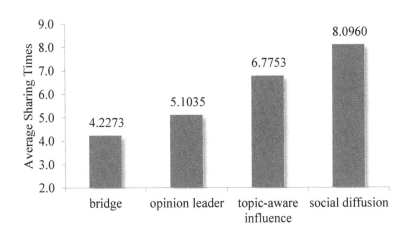

Fig. 3. The average sharing times of coupon in four diffusion strategies

5   Conclusion

Our proposed mechanism utilizes the theory of social influence and consumer psychology on improving coupon sharing intention. The mechanism was empirically verified by the experiments conducted on popular social networking site Facebook. The experimental results indicate our social coupon diffusion mechanism performs better than other three common used coupon diffusion approaches in term of average sharing rate, target coverage rate, and coupon download rate. This mechanism indeed can help marketers propagate a coupon to the right endorsers and the right customers wider and continuously.

References

1. Juniper Research: Mobile coupon users to pass 1 billion by 2019. Juniper research finds (2014). http://www.juniperresearch.com/press-release/coupons-pr1
2. Neilsen: Shopping Lists: How Mobile Helps Consumers Tick All the Boxes (2014). http://www.nielsen.com/us/en/insights/news/2014/shopping-lists-how-mobile-helps-consumers-tick-all-the-boxes.html
3. Nomadicpixie: Who is the Average Coupon User? (2011). http://www.afullcup.com/Grocery-Coupon-Blog/who-is-the-average-coupon-user/
4. Consumer Reports: Best coupon apps for grocery shopping: you don't have to be a coupon-clipping maven to save big (2013). http://www.consumerreports.org/cro/2013/08/best-coupon-apps/index.htm
5. Chandon, P., Wansink, B., Laurent, G.: A benefit congruency framework of sales promotion effectiveness. J. Mark. **64**(4), 65–81 (2000)
6. Heilman, C.M., Nakamoto, K., Rao, A.G.: Pleasant surprises: consumer response to unexpected in-store coupons. J. Mark. Res. **39**(2), 242–252 (2002)
7. Gfk: Digital Coupon Redeemer. Shopper Trends (2013)
8. Zhu, D.H., Chang, Y.P., Luo, J.J., Li, X.: Understanding the adoption of location-based recommendation agents among active users of social networking sites. Inf. Process. Manag. **50**(5), 675–682 (2014)
9. Banerjee, S.S., Dholakia, R.R.: Mobile advertising: does location based advertising work? Int. J. Mob. Mark. **3**(2), 68–75 (2008)
10. Jesse Holcomb, J.G., Mitchell, A.: News use across social media platforms. Pew Res. Cent. (2013)
11. Jin, J., Turner, S.J., Lee, B.S., Zhong, J., He, B.: HPC simulations of information propagation over social networks. Procedia Comput. Sci. **9**, 292–301 (2012)
12. Zhu, T., Wang, B., Wu, B., Zhu, C.: Maximizing the spread of influence ranking in social networks. Inf. Sci. **278**, 535–544 (2014)
13. Lee, M.K.O., Shi, N., Cheung, C.M.K., Lim, K.H., Sia, C.L.: Consumer's decision to shop online: the moderating role of positive informational social influence. Inf. Manag. **48**(6), 185–191 (2011)
14. Turner, J.C., Hogg, M.A., Oakes, P.J., Reicher, S.D., Wetherell, M.S.: Rediscovering the social group: a self-categorization theory. Contemp. Sociol. (1987)
15. Friedkin, N.E.: A Structural Theory of Social Influence. Cambridge University Press, Cambridge (2006)

16. Tierney, J.: Customers Want Online Shopping Options (2013). http://loyalty360.org/resources/article/customers-want-online-shopping-options. Accessed 14 Feb 2015
17. Yoo, C.W., Sanders, G.L., Moon, J.: Exploring the effect of e-WOM participation on e-Loyalty in e-commerce. Decis. Support Syst. **55**(3), 669–678 (2013)
18. Li, Y., Bandar, Z.A., McLean, D.: An approach for measuring semantic similarity between words using multiple information sources. IEEE Trans. Knowl. Data Eng. **15**(4), 871–882 (2003)
19. GSMA: Bridging the gender gap: Mobile access and usage in lowland middle-income countries (2015)

Effect of Instant Messenger Use on Purchase Decision of Consumers: The Role of Communication Quality and Content

Zhepeng Lu, Xiaoshan Wang, and Jinghua Huang$^{(\boxtimes)}$

School of Economics and Management, Tsinghua University,
Beijing 100084, China
huangjh@sem.tsinghua.edu.cn

Abstract. The uncertainty brought by the separation of information flow and product flow has become an important obstacle to the development of e-commerce. An increasing number of e-commerce websites have chosen to adopt an instant messenger (e.g., Wangwang) in the website as a communication tool between buyers and sellers to reduce this uncertainty. Based on uncertainty reduction theory, the relationship between the use of an instant messenger and the purchase decision of consumers is explored. Logit regression models were used to analyze the secondary data collected from a seller on Taobao.com, and the following findings are determined: (1) Customers who use Wangwang tend to purchase, and the effect of Wangwang use on purchase decisions is stronger for consumers who are indirect visitors. (2) The more quickly and more frequently sellers reply to questions, the more likely customers are to make purchase decisions. (3) Communicating about contents related to product quality in the conversation is positively associated with the consumer's purchase decision. (4) The negative effect of product fit-related contents to purchase decision will be mitigated if it is not the first conversation between consumers and sellers.

Keywords: Wangwang · Uncertainty · Uncertainty reduction · Communication · Purchase decision · Instant messenger

1 Introduction

Online shopping has become increasingly important for consumers with the development of the Internet. In 2014, the trading volumes of the online market in China reached 2,800 billion, which accounted for 10.7% of the total retail sales of consumer goods, and this amount will continue to increase in the following five years [1]. However, consumers cannot experience products themselves before purchasing when they shop online, thereby resulting in uncertainty and hampering purchase decision. Some shopping sites have launched an instant messenger (IM) to answer questions and thus reduce uncertainty. The IM service "Ali Wangwang" (Wangwang) provided by the well-known Chinese e-commerce site Taobao.com is a good example. A report from Morgan Stanley states that buyers and sellers may clarify their desire and demands through Wangwang, thereby enhancing the chances for online transactions [2].

© Springer International Publishing AG 2017
M. Fan et al. (Eds.): WeB 2016, LNBIP 296, pp. 125–138, 2017.
https://doi.org/10.1007/978-3-319-69644-7_13

The issue that needs to be addressed is whether or not IM usage has a certain influence on the purchase decision of consumers. If so, then how can IM be used to encourage purchasing behavior? Ou et al. claimed that the use of IM facilitates repeat transactions with sellers by building swift guanxi through interactivity and presence [3]. Kang et al. demonstrated that the use of IM positively affects the perceived interactivity of consumers, thereby increasing transaction intention [4]. The afore-mentioned study paid attention to the function of IM that it can help build a relationship between buyers and sellers. In addition, some scholars ascertained that IM can also facilitate communication, and the main objective of communication is uncertainty reduction [5–7]. Adjei et al. argued that communication quality among consumers can reduce uncertainty about a firm and its offerings, and sales can be influenced [8]; however, their study did not explore communication content.

In this study, we analyze whether the use of IM (represented by Wangwang) makes a difference in the purchase decision of consumers on the basis of uncertainty reduction theory with secondary data. Then, we discuss through communication quality and communication content how a purchase behavior is facilitated by communicating via Wangwang and also investigate the role of repeated dialogues as moderator between communication content and purchase decision.

2 Literature Review

2.1 Uncertainty of Online Shopping

Uncertainty of online shopping is the degree to which consumers fail to make predictions about a product or a firm with which they are dealing [9]. This uncertainty consists of seller uncertainty and product uncertainty [9–11]. Many scholars believe that seller uncertainty is mitigated by trust building mechanisms, such as feedback ratings and third-party escrows [10, 12]. Product uncertainty consists of product quality uncertainty and product fit uncertainty. Product quality uncertainty refers to the difficulty of consumers to evaluate product attributes and to predict how a product will perform in the future [10, 13, 14]. Product fit uncertainty is defined as the degree to which consumers cannot assess whether the attributes of a product match their preference [11]. The problem of product fit uncertainty becomes prominent because consumers cannot test products when they shop online.

Uncertainty reduction may move consumers close to a purchase decision. Pavlou et al. stated that perceived uncertainty may result in the probability of suffering a loss [9]. As a consequence, consumers reduce intentions to purchase and prevent the actual transaction behavior. Adjei et al. found that consumers in online brand communities buy products after uncertainty is reduced [8].

The aforementioned works, which discuss the classification and the influence of uncertainty on the purchase decision of consumers, provide a basis for us to explore how communication via Wangwang affects the purchase behavior of consumers by reducing product quality uncertainty and fit uncertainty.

2.2   Online Communication

Online communication is negligible for consumers when they purchase online. Communication can help buyers and sellers build relationships [3], positively influence the perceived interactivity of consumers [4], and finally enable consumers to make a purchase decision. Some researchers believe that communication can also positively affect the purchase behavior of consumers by reducing uncertainty [7, 8, 11].

The existing literature investigates how uncertainty is reduced mainly from the perspectives of communication quality and communication content. The factors that are most frequently mentioned as dimensions of communication quality are (1) timeliness, which refers to the speed of responses to the posted message [6, 7]; (2) frequency, which refers to the number of responses to the posted messages [15, 16]; and (3) duration/length of encounter, which refers to the number of words in the selected thread [8, 17]. Adjei et al. claimed that communication quality among consumers in an online brand community can significantly reduce product (service) uncertainty [8].

In addition to communication quality, the effect of communication content, which consists of product quality- and product fit-related contents, has been analyzed. Hong and Pavlou believed that consumers can obtain a deep understanding of whether a product fits their preference from the experience of other consumers [11]. Consumers can also validate the existing seller-provided information from other consumers, and additional information about product quality can be obtained.

In this study, we are concerned about the role of communication in reducing uncertainty instead of its role in building a relationship [3]. We refer to the dimensions of communication quality and communication content to explore the effect of communication via Wangwang on the purchase decision of consumers.

2.3   Summary of Existing Problems

(1) The majority of studies investigated the role of communication in promoting perceived interactivity and building relationship when analyzing the influence of communication on the purchase decision of consumers [3, 4]. One of the main objectives of communication is uncertainty reduction [5–7]. However, there are few studies discussing the effect of IM on the purchase decision of consumers through uncertainty reduction.

(2) Adjei et al. analyzed the effect of communication on the purchase behavior of consumers on the basis of uncertainty reduction theory [8]. However, they discussed communication quality only, and adequate analysis of communication content was not incorporated.

3   Research Hypotheses

3.1   Effect of Wangwang Use on the Purchase Decision of Consumers

The use of Wangwang can help consumers to reduce information asymmetry, thereby reducing uncertainty. As information seekers, consumers acquire significant information by communicating with sellers using Wangwang, thereby reducing a large set of possibilities about sellers and their products to a small set; hence, uncertainty is eventually reduced. The level of uncertainty is negatively associated with consumer intention and actual online purchases [9]. Therefore, we propose the following:

H1: The use of Wangwang is positively associated with the purchase decision of consumers.

Some scholars found that information demanded to make purchase decision for different types of consumers is different [25]. We extend the study of Tong et al. [26], and define that direct visitors are consumers directed by links in shopping favorite lists, shopping collections, purchase history in the shopping platform to access the store. These consumers have once visited the store and they have more knowledge about the store and its products [27, 28]. Indirect visitors are consumers accessing the store through searching the related products or clicking banner advertisement links in the platform. They have no shopping experience in this store and are unfamiliar about the quality of the products, which means they are in high level of uncertainty. Hence, before making any purchase decision, indirect visitors need to collect more information compared to direct visitors. By communicating with sellers through Wangwang, consumers can get specific answers to their questions quickly and thus reduce uncertainty. Therefore, we propose the following:

H2: The impact of use of Wangwang on the purchase decision will be greater if the consumer is an indirect visitor.

3.2   Effect of Communication Quality and Content in IM on the Purchase Decision of Consumers

According to the studies that discuss uncertainty reduction through communication [5, 6], communication quality [17–19] and communication content [8, 11] may reduce uncertainty when consumers shop online, thereby bringing them close to a purchase decision.

From the perspective of communication quality, a prompt response from sellers helps information seekers to immediately acquire a comprehensive understanding of products and to clarify their goals [7], thereby attenuating the negative effect of uncertainty. More frequent responses from sellers correspond to more information related to product quality and service that consumers can acquire as information seekers. A long duration/length of encounter may imply that descriptive and in-depth information can be provided to consumers [7]. This additional information can help mitigate the information asymmetry between consumers and sellers/products, thereby reducing uncertainty and leading to a purchase decision [9]. Therefore, we propose the following:

H3a: In communication via Wangwang, timely reply from sellers is positively associated with the purchase decision of consumers.

H3b: In communication via Wangwang, the number of replies from sellers is positively associated with the purchase decision of consumers.

H3c: In communication via Wangwang, the length of encounter is positively associated with the purchase decision of consumers.

The product quality- and product fit-related contents mentioned in the communication via Wangwang may help to reduce the uncertainty level. According to the definition of product quality uncertainty [10, 14], we define that product quality-related contents are the information related to product attributes, such as product condition and quality. In addition to the information in a website that displays products and feedback ratings, the detailed information provided by sellers about product attributes according to the questions of consumers can help consumers gain an understanding of product quality. This detailed information leads to an accurate prediction about the performance of a product in the future and to the reduction of product quality uncertainty [10]. We extend the study of Hong and Pavlou [11], and define product fit-related contents as the information that helps consumers determine whether the attributes of a product match their preference. Sellers can recommend products to consumers to match their preference according to previous consumer feedback. The communication with sellers in Wangwang helps consumers determine the fit between product attributes and their own preferences, thereby reducing product fit uncertainty [11]. The uncertainty reduction raises the intention of consumers to purchase, thereby leading to an actual purchase decision. Therefore, we propose the following:

H4a: If a product quality-related content is mentioned in communication via Wangwang, then consumers tend to make a purchase decision.

H4b: If a product fit-related content is mentioned in communication via Wangwang, then consumers tend to make a purchase decision.

For consumers that are the first time to communicate with sellers and consumers that have already had conversations with sellers for several times, the effect of communication content on purchase decision may be different. We define dialogues between sellers and consumers who have already communicated with sellers before as repeated dialogues. When consumers start the conversation about product quality- or fit-related contents with sellers in the repeated dialogues, sellers get additional opportunity to provide more product features, or get to know consumers individual preference in details and advise customized solutions in response. Therefore, together with information obtained from the previous dialogue, consumers will get more comprehensive understanding about product quality and the fit between the product and their preference compared to consumers without repeated dialogues. Therefore, we propose the following:

H5a: The impact of mentioning product quality-related contents via Wangwang on the purchase decision will be greater if the dialogue is a repeated dialogue.

H5b: The impact of mentioning product fit-related contents via Wangwang on the purchase decision will be greater if the dialogue is a repeated dialogue.

4 Research Methodology

4.1 Data Collection

Data from a skin care product store in Taobao.com were collected. We finally obtained the secondary data, which include 33,998 browsing records of visitors (only the PC-side data were available on Taobao.com), 4,778 purchase records, 19,610 reviews for products, 7,603 Wangwang dialogues (111,754 postings), 5396 consumer characteristics and 235 product display pages, between June 1, 2015 and December 8, 2015.

4.2 Key Variables

The browsing record for each visitor had no real Taobao ID (only randomly assigned IDs were available) because of the privacy policy of Taobao.com. Thus, we could not learn from the raw data whether a visitor in the store used Wangwang (variable *Wangwang* took a value of 1 if the visitor used Wangwang; otherwise, 0 was taken) or purchased a product (variable *Purchase* took a value of 1 if the visitor made a purchase decision; otherwise, 0 was taken). We matched the timestamp of the Wangwang dialogue and the product link mentioned in the Wangwang dialogue with the time of visit and with the product display page in the browsing records, respectively, to confirm whether a visitor used Wangwang. We matched the address in the purchase records, the created time of purchase record, and the products purchased with the IP location of visitors, with the time of visit and with the product display page in the browsing records, respectively, to determine whether a visitor made a purchase decision. *Type* refers to the type of consumers. *Type* took a value of 1 if the consumer was a direct visitor (the link in the favorite list or purchase history was used to access this store), whereas a value of 0 was taken if the consumer was an indirect visitor (this store was accessed by searching the related products or other advertisement links). The data between June 1, 2015 and June 30, 2015 were analyzed. A total of 4,069 browsing records in this period were matched with 156 purchase records and 139 Wangwang dialogues.

In addition to the main variables in the first and second hypothesis, we used several control variables associated with browse behavior [20–22]. *PageWord* refers to the average number of words. *PagePicture* refers to the average number of pictures. *ProductReview* refers to the average number of reviews. *Price* refers to the average number of prices in product pages that a visitor browsed.

We analyzed the content of Wangwang dialogues to obtain the following indicators. *Speed* refers to the time interval between the first posting of consumers and the reply of sellers. *Frequency* refers to the number of replies from sellers to the postings of consumers. *Length* refers to the number of words in the Wangwang dialogue. *RptDialogue* refers to whether the dialogue happened between sellers and a consumer who has already communicated with sellers before. It took a value of 1 if the consumer once has a conversation with the seller; otherwise, 0 was taken.

We obtained the value of variables about communication content in the following procedures: First, we randomly selected 437 dialogues, which contained 6,476 postings. Second, we manually coded these 6,476 postings into several categories. Service quality (e.g., product return, gifts, express policy) may have an effect on the purchase decision of consumers [23]; thus, we decided to consider service quality in the classification. Each posting was classified into one of four groups: product quality, product fit, service quality-related contents, and others. Third, we trained the algorithms of support vector machine (SVM) and naive Bayes (NB) using 89.6% (i.e., 5800) of the coded postings as input to build classification models. Fourth, we used models to classify the remaining 10.4% (i.e., 676) of the postings. The accuracy level of the SVM was 82.98%, which was higher than that of NB (78.01%). Finally, the SVM was applied to classify the remaining 105,278 postings into four groups. *Quality* refers to product quality-related contents, and it takes a value of 1 as long as one posting in the Wangwang dialogue was classified into the product quality-related contents group; otherwise, a value of 0 was taken. *Fit* and *Service* refer to product fit and service quality-related contents, respectively, and their definitions were similar to *Quality*. We also controlled for consumers characteristics. *CreditLevel* refers to consumers' credit evaluated by sellers. *MemberLevel* refers to consumers' membership rank, which is associated with their total expenditure in Taobao.com. *ActiveLevel* refers to consumers' activeness in Wangwang, which is associated with the time spent in Wangwang dialogues since they signed up. Username was used to match the records of Wangwang dialogues and consumer characteristics. Some consumers may have changed their username or closed accounts, so there are 312 Wangwang dialogues cannot be matched. Finally, we used the rest of Wangwang dialogues (7,291 dialogues, 104,858 postings) to perform the empirical tests.

4.3 Model Specification

The dependent variable Purchase represented binary variables; thus, we applied logit model to estimate the equation. Equation (1) was used to test hypothesis 1 and 2 by determining the effect of Wangwang use on the purchase decision of consumers, and the moderation effect of consumer type, controlling for their type, average words, average pictures, average product reviews, and average price for the pages they browsed. Equation (2) was used to test hypotheses 2a to 2c and 3a to 3b by evaluating the effect of the speed and frequency of replies, length of encounter, product quality, and product fit-related contents on the purchase decision of consumers, controlling for service quality-related contents and consumers' characteristics. We also added variable *RptDialogue* and the interaction term *RptDialogue * Fit* and *RptDialogue * Quality* to Eq. (2) to test hypotheses 5a to 5b. ϵ was the random error.

$$
\ln\left(\frac{P(Purchase = 1)}{1 - P(Purchase = 1)}\right) = \beta_0 + \beta_1 Wangwang + \beta_2 {*} Type + \beta_3 * Type * Wangwang
$$
$$
+ \beta_4 {*} PageWord + \beta_5 {*} PagePicture + \beta_6 {*} ProductReview
$$
$$
+ \beta_7 {*} Price + \epsilon \tag{1}
$$

$$\ln\left(\frac{P(Purchase = 1)}{1 - P(Purchase = 1)}\right) = \beta_0 + \beta_1 * Speed + \beta_2 * Frequency + \beta_3 * Length$$
$$+ \beta_4 * Fit + \beta_5 * Quality + \beta_6 * Service + \beta_7 * CreditLevel$$
$$+ \beta_8 * MembershipLevel + \beta_9 * ActiveLevel + \epsilon$$

$$(2)$$

5   Data Analysis and Results

5.1   Descriptive Statistics

The data between June 1, 2015 and June 30, 2015 were used to estimate model 1. Table 1 shows that only 3.8% of 4,069 visitors purchase a product. This finding implies that the purchase conversion rate is low online. In total, 3.4% of visitors use Wangwang.

The descriptive statistics of 7,291 Wangwang dialogues in Table 2 show that 2,390 purchase records exist after dialogues. Sellers reply 7.4 times and give the first reply in 692.8 s in each dialogue on average. Of all the dialogues, 26.0% mention product fit-related contents, whereas 56.7% mention product quality-related contents.

Table 1. Descriptive statistics for all the consumers

Variable	Mean	Std. Dev.	Min	Max
Purchase	0.038	0.192	0	1
Wangwang	0.034	0.182	0	1
Type	0.251	0.433	0	1
PagePicture	7.923	3.952	0	18
PageWord	1,299.684	658.056	0	3711
ProductReview	27.841	23.147	0	130
Price	109.409	105.730	0	666

5.2   Hypotheses Testing

(1)   Effect of Wangwang use on purchase decision:

The estimated results for model 1 are reposted in Table 3. The use of Wangwang has a positive coefficient (b = 3.341, p < 0.01), and the marginal effect of Wangwang on purchase decision is 0.376 (using the mfx command in Stata), i.e., the use of Wangwang is associated with a 0.376 increase in the probability of the purchase decision of consumers compared with when Wangwang is not used. Therefore, H1 is supported. The empirical results in model 2 indicate that the interaction term between use of Wangwang and consumer type is negatively significant (b = −1.096, p < 0.05), suggesting that consumer type acts as an moderator on the relationship between use of

Table 2. Descriptive statistics variables related to Wangwang dialogues

Variable	Mean	Std. Dev.	Min	Max
Purchase	0.328	0.469	0	1
Frequency	7.415	9.462	0	122
Speed	692.801	3744.953	0	86400
Length	345.648	365.577	2	5088
Quality	0.567	0.496	0	1
Fit	0.260	0.439	0	1
RptDialogue	0.260	0.439	0	1
Service	0.599	0.490	0	1
CreditLevel	5.938	1.660	1	13
MemberLevel	3.038	0.955	0	6
ActiveLevel	10628.580	20296.320	0	214423

Table 3. Estimation results for effect of Wangwang use on purchase decision

IV	Model 1		Model 2	
	b	se	b	se
Intercept	-4.102***	0.313	-4.769***	0.330
Wangwang	3.341***	0.206	3.791***	0.277
Type			1.528***	0.208
Wangwang*Type			-1.096**	0.425
Control variables	–	–	–	–
PagePicture	-0.082***	0.027	-0.082**	0.027
PageWord	4.27E-04***	1.32E-02	4.23E-04***	1.35E-04
ProductReview	0.007*	0.004	0.009**	0.004
Price	0.002**	0.001	0.002***	0.001
Log-likelihood	-544.204		-516.887	
N	4069		4069	
Pseudo R2(%)	17.76		21.89	

$*p < 0.1; **p < 0.05; ***p < 0.01$

Wangwang and consumers' purchase decision. Therefore, H2 is supported. Direct visitors have more knowledge about the store and its products, so the information obtained through communication via Wangwang may be trival, thus it will have weaker influence on reducing uncertainty compared to indirect visitors.

Of all the control variables in both model 1 and model 2, the average number of pictures in pages that consumers browsed has significant negative effects on their purchase decision. This result was obtained because skin care products sold in a sample store are typical experience products whose attributes cannot be adequately conveyed with pictures; thus, uncertainty cannot be reduced. Many ineffective pictures decrease the attraction of the pages to consumers, thereby resulting in a low possibility to make a purchase decision. As the result suggested, price of the browsed products impacts consumers' purchase decision positively. This may be due to large quantities of low price counterfeit goods in Taobao.com. Consumes are more likely to treat products with low price as counterfeit goods. The effect of other control variables is similar to that discussed in the previous literature.

(2) **Effect of communication through Wangwang on purchase decision:**

The estimation results for model 3 are reposted in Table 4. According to the perspective of communication quality, frequency of seller replies positively influences the purchase decision of consumers (b = 0.073, p < 0.01). The marginal effect is 0.015, which indicates that one more response from a seller to a consumer is associated with a 0.015 increase in the possibility of the purchase decision of the consumer. Therefore, H3b is supported. We also determine that the speed of the first reply has significant positive effects on the purchase decision of consumers. Therefore, H3a is supported. However, the effect of the length of encounter is significantly negative because the increase in the number of words in the dialogue indicates that a consumer may have many questions about products. This scenario implies a high level of uncertainty. Consequently, consumers do not make a purchase decision.

The estimation results in model 3 show that of the two communication content variables, product quality-related contents have more significant effects on purchase decision. Therefore, H4a is supported. However, the relationship between the purchase

Table 4. Estimation results for the effect of communication quality and contents

IV	Model 3		Model 4	
	b	se	b	se
Intercept	−2.569***	0.136	−2.467***	0.138
Frequency	0.073***	0.006	0.070***	0.006
Speed	−3.770E-05***	6.090E-06	−3.74E-05***	6.06E-06
Length	−0.002***	1.629E-04	−0.001***	1.64E-04
Quality	0.378***	0.062	0.316***	0.069
Quality*RptDialogue			0.020	0.134
Fit	-0.138*	0.072	−0.232***	0.079
Fit*RptDialogue			0.388**	0.164
RptDialogue			−0.483***	0.093
Control variables	–	–	–	–
Service	1.559***	0.067	1.559***	0.067
CreditLevel	0.018	0.020	0.023	0.020
MemberLevel	0.172***	0.035	0.174***	0.035
ActiveLevel	3.150E-06**	1.530E-06	3.32E-06**	1.54E-06
Log-likelihood	−4022.790		−4001.101	
N	7291		7291	
Pseudo R2(%)	12.78		13.25	

*p < 0.1; **p < 0.05; ***p < 0.01

decision of consumers and product fit-related contents is negatively significant at the level of 10%. On the one hand, this scenario is observed probably because the information on the other consumers' experience provided by sellers cannot reduce the uncertainty when consumers strongly doubt the level of fit between the product and their preference. On the other hand, sellers may recommend other products to consumers according to consumer preference; however, these recommended products are unfamiliar to consumers, and consumer uncertainty is increased. Therefore, consumers experience difficulty in making a purchase decision.

As the results of model 4 in Table 4 indicate, consumers who have repeated dialogues with sellers will reduce the probability of purchasing (b = −0.483, p < .01). The interaction term between product quality-related contents and repeated dialogues is not significant, which indicates that after first conversation, communicating with sellers again and again will not reduce product quality uncertainty of consumers any more. Product quality-related contents are contents about product attributes such as product appearance, expiry date, ingredients. These contents are objective and easy to obtain through product webpages or communicating with sellers. Once consumers acquire information about these contents, talking about them again will not provide additional information that may help to reduce uncertainty and make purchase decision. Therefore, H5a is not supported. The empirical results in model 4 suggest that repeated dialogues can mitigate the negative relationship between product fit-related contents and consumers' purchase decision (b = 0.388, p < .05). Unlike product quality-related contents, product fit-related contents are more subjective and if there are chances to hold more conversations, sellers are more likely to convince consumers about the fit by using more relevant examples and detailed description.

6 Summary and Conclusion

6.1 Key Findings

This study analyzes the relationship between the use of IM, which is represented by Wangwang, and the purchase decision of consumers by using the data from a skin care product store. Our results show that consumers who use Wangwang tend to make a purchase, and the effect of Wangwang use on purchase decisions is stronger for consumers who are indirect visitors. During communication via Wangwang, fast and frequent replies from sellers increase the likelihood that consumers will make purchase decisions. Communication content related to product quality has positive effects on the purchase decision of consumers, and the negative effect of product fit-related contents to purchase decision will be mitigated if it is not the first conversation between consumers and sellers.

This study contributes to the literature on consumer behavior in the following points:

First, our research is different from the previous literature on the role of communication in promoting perceived interactivity and in building relationships to influence purchase decision [3, 4]. We analyze the effect of communication between consumers and sellers on the basis of the contribution of communication in reducing uncertainty. This effect may help in understanding the antecedents of online shopping decision from a new perspective.

Second, secondary data are used for the first time to analyze the communication in Wangwang dialogues through the cooperation of a store on Taobao.com.

Third, in addition to communication quality, we analyze the content of communication based on the classification of uncertainty in marketing literature into product quality- and product fit-related contents. The effect of communication content is explored, thereby providing evidence on uncertainty reduction theory.

6.2   Managerial Implications

This study has the following managerial implications:

First, communication between consumers and sellers is of vital importance, especially for consumers who have never been to the store before. Sellers could take advantages of monitoring tools such as Shengyicanmou provided by Taobao.com or other third party companies to identify a first visit, and let an experienced staff communicate with the consumer.

Second, sellers should reply to consumers as quickly as possible because a quick reply can provide information in time before the uncertainty results in a negative effect. Moreover, a quick response can leave consumers with a good impression of high-quality service, thereby leading to a pleasant online shopping experience.

Third, sellers can convey rich and comprehensive information to consumers by replying frequently, thereby helping to reduce uncertainty and turn information seekers into consumers.

Fourth, sellers should realize that they still need other communication tools (e.g., online video) to help consumers make decisions by providing detailed information because communication via Wangwang can effectively reduce product quality uncertainty but not product fit uncertainty.

Last but not least, sellers can give more contents about the fit between products and a consumer's preference when the consumer starts several conversations. A conversation after another indicates that the consumer still considers purchasing the product seriously. More product fit-related contents with relevant examples and detailed description may look trustworthy, and thus the consumer is likely to be convinced that this product matches his/her preference well.

6.3   Limitations

First, only PC-side browsing records in the store are available because of the privacy policy of Taobao.com. Thus, we do not have mobile device-side data to estimate model 1. The decision of consumers may be influenced by the difference (e.g., screen size) between a PC and a mobile device. Thus, the estimated results may not match our results.

Second, the demographic information of consumers such as age, gender is not controlled when we analyze the relationship between communication and purchase decision (i.e., model 2). The demographic information may play a part in making a purchase decision and may change the estimation results.

Third, the data analyzed in this study come from a store that sells skin care products, which are typical experience products [24]. Consumers may rely on different kinds of information when they buy other products. Thus, the findings of this study may not be applicable in other contexts.

References

1. IResearch, Online markets in China maintain a stable and rapid growth in 2014 (2014). http://www.iresearch.com.cn/view/245910.html
2. Ji, R., Meeker, M.: Creating consumer value in digital China. Morgan Stanley (2005)
3. Ou, C.X., Pavlou, P.A., Davidson, R.M.: Swift guanxi in online marketplaces: the role of computer-mediated communication technologies. MIS Q. **38**(1), 209–230 (2014)
4. Kang, L., Wang, X., Tan, C.H., Zhao, J.L.: Understanding the antecedents and consequences of live chat use in electronic markets. J. Organ. Comput. Electron. Commer. **25**(2), 117–139 (2015)
5. Carlson, J.R., Zmud, R.W.: Channel expansion theory and the experiential nature of media richness perceptions. Acad. Manag. J. **42**(2), 153–170 (1999)
6. Daft, R.L., Lengel, R.H.: Information richness: a new approach to manager information processing and organization design. In: Staw, B.M., Cummings, L.L. (eds.) Research in Organization Behavior, vol. 6, pp. 191–233. JAI, Greenwich (1984)
7. Weiss, A.M., Lurie, N.H., MacInnis, D.J.: Listening to strangers: whose responses are valuable, how valuable are they, and why? J. Mark. Res. **45**(4), 425–436 (2008)
8. Adjei, M.T., Noble, S.M., Noble, C.H.: The influence of c2c communications in online brand communities on customer purchase behavior. J. Acad. Mark. Sci. **38**(5), 634–653 (2010)
9. Pavlou, P.A., Liang, H., Xue, Y.: Understanding and mitigating uncertainty in online exchange relationships: a principal-agent perspective. MIS Q. **31**(1), 105–136 (2007)
10. Dimoka, A., Hong, Y., Pavlou, P.A.: On product uncertainty in online markets: theory and evidence. MIS Q. **32**(2), 395–426 (2012)
11. Hong, Y., Pavlou, P.A.: Product fit uncertainty in online markets: nature, effects and antecedents. Inf. Syst. Res. **25**(2), 328–344 (2014)
12. Benbasat, I., Gefen, D., Pavlou, P.A.: Special issue: trust in online environments. J. Manag. Information Syst. **24**(4), 5–11 (2008)
13. Arrow, K.J.: Uncertainty and the welfare economics of medical care. Am. Econ. Rev. **53**(5), 941–973 (1963)
14. Ghose, A.: Internet exchanges for used goods: an empirical analysis of trade patterns and adverse selection. MIS Q. **33**(2), 263–291 (2009)
15. Doney, P.M., Cannon, J.P.: An examination of the nature of trust in buyer-seller relationships. J. Mark. **61**(2), 35–51 (1997)
16. Berger, C.R.: Communicating under uncertainty. In: Rolloff, M.E., Miller, G.R. (eds.) Interpersonal Processes: New Directions in Communication Research, pp. 39–62. Sage, Newbury Park (1987)
17. Berger, C.R., Calabrese, R.J.: Some explorations in initial interaction and beyond: toward a developmental theory of interpersonal communication. Hum. Commun. Res. **1**(2), 99–112 (1975)

18. Mohr, J., Fisher, R., Nevin, J.R.: Collaborative communication in interfirm relationships: moderating effects of integration and control. J. Mark. **60**(3), 103–115 (1996)
19. Mohr, J., Spekman, R.: Characteristics of partnership success: partnership attributes, communication behavior and conflict resolution techniques. Strateg. Manag. J. **15**(2), 135–152 (1994)
20. Li, T., Ye, Q., Li, Y.: An investigation regarding stickiness & purchase conversion from the perspective of traffic source. China J. Inf. Syst. **2**, 46–53 (2012)
21. Chiang, K.P., Dholakia, R.R.: Factors driving consumer intention to shop online: an empirical investigation. J. Consum. Psychol. **13**(1), 177–183 (2003)
22. Abbasi, A., Chen, H.: Cybergate: a design framework and system for text analysis of computer-mediated communication. MIS Q. **32**(4), 811–837 (2008)
23. Valvi, A.C., Fragkos, K.C.: Critical review of the e-loyalty literature: a purchase-centred framework. Electron. Commer. Res. **12**(3), 331–378 (2012)
24. Nelson, P.: Advertising as information. J. Polit. Econ. **82**(4), 729–754 (1974)
25. Qin, E., Lu, X.: The initial buyers and repeat buyers in online retailing: an empirical Study on Taobao.com. Econ. Manag. J. **34**(6), 82–90 (2012)
26. Li, T., Ye, Q., Li, Y.: An investigation regarding stickiness and purchase conversion from the perspective of traffic source. China J. Inf. Syst. **1**, 46–53 (2012)
27. Meyer, R.J.: A descriptive model of consumer information search behavior. Mark. Sci. **1**(1), 93–121 (1982)
28. Reilly, M.D., Conover, J.N.: Meta -analysis: integrating results from consumer research studies. Adv. Consum. Res. **10**(4), 509–513 (1997)

Booking High-Complex Travel Products on the Internet: The Role of Trust, Convenience, and Attitude

Maria Madlberger[(✉)]

Webster Vienna Private University, Praterstraße 23, 1020 Vienna, Austria
maria.madlberger@webster.ac.at

Abstract. The number of electronic channels in the tourism industry has significantly increased and led to substantial changes in consumers' booking behavior. The majority of extant research in this area addresses single travel products that show a low level of complexity (e.g., airline tickets). This study focuses on the role of product complexity by testing a research model on online booking intention of high-complex travel products which have been less investigated. The results show that for high-complex travel products, trust shows a stronger positive impact on attitude towards online booking compared to previous research. In contrast, perceived convenience of the Internet as a booking channel shows no significant impact which differs from previous findings on low-complex travel products. Attitude and subjective norm both positively influence online booking intention. The results suggest that the role of product complexity in drivers of online purchase intention should be revisited and its potential moderating impact should be considered.

Keywords: Online booking intention · Trust · Convenience · Product complexity · Tourism industry

1 Introduction

Few industries have experienced a similar transformation through Internet-based distribution as the tourism sector. The number of distribution channels in this industry has significantly increased. Whereas travelers primarily interacted with traditional travel agencies and tour operators prior to the availability of the Internet, online booking channels nowadays have become a major distribution channel [1]. In particular, global online third-party platforms established competition with traditional travel agencies [2]. At present, consumers use multiple channels if they decide on booking trips [3]. Recent figures show the sharp increase of online and mobile channels at cost of traditional offline channels. In Germany, the percentage of travelers who make travel bookings on the Internet has increased from 28% to 69% during the time span from 2006 to 2016. Similarly, the percentage of users who book via online travel agencies rose from 18% to 55%. In contrast, the portion of users who book via physical travel agencies decreased from 61% to 34% within ten years [4]. In the UK, the percentage of users who book travel products online increased to 79%, 50% use online travel agencies, 47% book via airline or hotel websites, and 19% use physical travel agencies for travel booking [5].

© Springer International Publishing AG 2017
M. Fan et al. (Eds.): WeB 2016, LNBIP 296, pp. 139–149, 2017.
https://doi.org/10.1007/978-3-319-69644-7_14

In the light of these developments, an understanding of what drives consumers' selection of travel booking channels is important. Whereas e-commerce and information systems research intensively investigated antecedents of online booking behavior in general, still little is known about the role of product characteristics which are a key factor in e-commerce. The majority of studies on antecedents of online travel booking do not differentiate between various travel products and conduct research in the context of single travel services, such as airline tickets or hotel reservations (e.g., [6–9]). Several studies address the issue of product complexity in the tourism industry [10–13] and conclude that complexity of travel products does not negatively affect intention or successful completion of online booking. However, research in this area is still fragmented.

The motivation of the study at hand is gaining first insights into Internet users' drivers to select online channels for booking high-complex travel products as discussed by [11]. These authors differentiate between products of low complexity, such as flights, hotel rooms, and car rental and products of high complexity, e.g., land-based vacations, cruises or tours. We seek to answer the following research question: *What are drivers of Internet users' intention to book high-complex travel products online?*

For this purpose, we developed a research model based on findings of research in e-commerce and online tourism. Theory-wise, the model is built upon the theory of planned behavior [14]. The research model is applied to online booking intention of a high-complex travel product, concretely organized circular tours. Whereas booking airline tickets or hotel rooms refers to one single service component, circular tours are a bundle of different service components [15, 16] including flight, hotels, bus or rental car, meals etc. The research model is tested with survey data. The results show that the majority of the hypotheses are supported, yet some findings differ from extant research in the context of low-complex travel products.

2 Literature Review

The tourism industry is a sector that has received major attention in information systems and e-commerce research. Travel products are strongly information-based, meaning that the purchase decision is solely based on information whereas the consumption takes place after the purchase at a different location [17]. Resulting from the key role of information, the travel business has become affected by disintermediation [18] once consumers were given the opportunity to access information on electronic reservation systems online. Since then, the number of available booking channels has largely increased and now encompasses online channels such as online travel agencies, travel search engines, social media, or travel companies' own websites in addition to existing offline channels [5].

When it comes to the analysis of online booking channels and their drivers, several antecedents have been identified. In an extensive literature review, [19] discuss 13 different antecedents of online purchasing, grouped into consumer, channel, and product characteristics. Product characteristics encompass attributes such as price, reviews, and variety whereas the channel-related factors refer to information and service quality, trust, payment, and customer relationship [20]. Both product-related and channel-related factors have an impact on purchase intention that itself is driven by intention to search for information in the particular channel [20]. Several studies investigate influence factors of consumers' perceptions and attitudinal beliefs, i.e., assessments of the object in question, as drivers of online booking behavior. Attributes of online travel agencies that are relevant to consumers are Web features such as booking flexibility, full transaction phase support, or sorting options, user friendliness and security as well as the possibility to find low fares [21]. Despite the advantages of online booking, offline travel agencies enjoy perceived benefits such as the knowledge and experience of the personnel, their friendliness, personal advice [22], helpfulness, favorable prices, and the quality of holiday packages [23]. [24] show that consumers consider attributes of offline and online travel agencies almost identically important. Also situational factors influence the choice of the booking channel. For example, groups of individuals rather choose a travel agency over the Internet than single individuals [25, 26]. Consumers who are in time pressure tend to visit travel agencies or consult brochures to speed up the booking process [3]. In the context of high-risk services, however, consumers switch from online information search to offline booking [27].

Studies that address the role of product complexity in the online tourism setting suggest that a high complexity of travel products does not pose a hurdle to online booking. In an experimental setting, it could not be shown that online booking of high-complex travel products causes more problems or booking errors than low-complex travel products [10]. Considering travel product complexity an independent variable, [12] finds no significant impact of this factor on online booking intention. A study that compares a high-complex travel product with a low-complex service product from the non-tourism context shows that, different from the primary assumption, online purchase intention is even more likely for the high-complex travel product [13]. [11] draw on travel purchase motivation and show that high-complex travel products differ from low-complex ones in that the former are associated with high informational motivation (i.e., higher need for provision of detailed information) whereas the latter are related to transactional motivation (i.e., completing the booking process).

3   Research Model

A large number of studies on drivers of consumers' acceptance of online shopping sites identifies trust as a key impact factor that shows a stronger impact on purchase intention than for example perceptions of the price [28]. As [29] show, trust directly and indirectly affects purchase intention by influencing attitude, which is in line with the theory of planned behavior [14]. Also in the context of the tourism industry, the important role of trust is demonstrated in various studies. Consumer trust in travel products and online channels influences intentions to book online [30] as well as loyalty [31]. As [32] show, trust is an important mediator that explains the indirect effects of transaction security, navigation functionality, and cost effectiveness on repurchase intention in the tourism industry. We therefore suggest:

H1: Trust in the Internet as a booking channel positively impacts attitude towards booking high-complex travel products on the Internet.

The Internet is particularly discussed as a channel that offers a maximum extent of convenience to consumers. As research has shown, convenience plays a major role in the formation of satisfaction with online shops and purchasing online [33, 34]. Studies also show the strong impact of convenience on online purchase intention [35, 36]. The importance of convenience for attitude towards online purchasing is also demonstrated by studies in the context of travel products [9] where convenience is one of the major drivers of customer satisfaction [7, 8]. Therefore we hypothesize:

H2: Perceived convenience of the Internet as a booking channel positively impacts attitude towards booking high-complex travel products on the Internet.

Numerous studies on e-commerce draw on the theory of planned behavior [14] and investigate attitude towards online purchasing as an antecedent of online purchase intention. Also in the tourism context attitude towards online booking has been confirmed as a driver of online booking intention [8, 9, 37, 38]. Thus we propose:

H3: Attitude towards booking high-complex travel products on the Internet positively impacts intention to book high-complex travel products on the Internet.

Subjective norm is another established antecedent of behavioral intention according to the theory of planned behavior [14]. In e-commerce research, subjective norm has been investigated in numerous studies, achieving mixed results. Whereas some studies show a significant impact of subjective norm (e.g., [39, 40]), others cannot confirm this effect (e.g., [41, 42]). Also in the travel sector, research does not always show a significant impact of subjective norm on online booking intention. Whereas [43] show that the impact of subjective norm is contingent upon the cultural background, [44] demonstrate a significant impact on the intention to stay loyal with an online reservation system. Seeking to clarify the contradiction in the extant literature and assuming a stronger role of the social context in high-complex travel products we hypothesize a significant impact of subjective norm:

H4: Subjective norm on the Internet as a booking channel positively impacts intention to book high-complex travel products on the Internet.

Figure 1 summarizes the research model graphically.

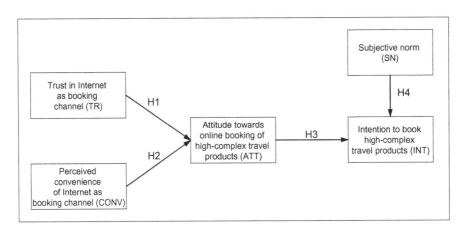

Fig. 1. Research model

4   Research Methodology

4.1   Measurement Items

All constructs included in the research model were measured with established multi-item scales that were adapted from the literature and applied five-point Likert scales where 1 stands for "totally agree" and 5 stands for "totally disagree" (except for the attitude items where answer categories ranged from e.g., "good idea" to "bad idea"). Three constructs were built based on three measurement items; subjective norm was measured with four items. The items for measuring convenience of booking high-complex travel products online were adapted from [45] by referring to booking of circular tours. The items measuring trust in the Internet as a channel for booking high-complex travel products were adapted from [46, 47]. The items related to subjective norm were adapted from the seminal paper by [48]. Finally intention to book high-complex travel products was measured by the modified scale of [49] that was applied to organized circular tours.

4.2   Data Collection

The research design consists of a survey. Data collection took place by standardized personal interviews with undergraduate and graduate students studying at an Austrian university. Whereas student samples may be criticized for lack of representativeness and potential resulting bias [50], this convenience sampling approach was chosen for several reasons. First, this study seeks to gain first insights into the overall issue of product complexity within a homogenous sample. Second, students are one of the population groups with highest Internet usage intensity [51, 52]. Third, young people are a large and quickly growing customer segment in the tourism industry that has reached a global share of 20% [53]. Thus, like in other studies, the student sample can be considered appropriate for the given purpose [54]. Prior to the survey a pre-test with 15 students was conducted, which resulted in minor changes in wording and question order. In the survey 136 questionnaires were completed. After removing questionnaires with incomplete answers, 122 questionnaires were used for data analysis.

The sample consists of 48.4% males and 51.6% females. The average age is 21.95 years with a standard deviation of 2.34 years. The majority (59.6%) is not employed, 31.2% are half-time employed, and 9.2% are fully employed. 85.2% study on the bachelor level. A filter question assured that all respondents have traveled at least once within the last 12 months from the time of the survey.

5   Results

5.1   Measurement Model

The test of the measurement model includes the analysis of consistency (Cronbach alpha, composite reliability), convergent, and discriminant validity. All factor loadings exceed the value of 0.5. All Cronbach alpha and composite reliability values are higher than the recommended value of 0.7 [55]. The average variance extracted (AVE) is higher than the recommended value of 0.5 [56]. The correlations among the latent variables are higher than the interconstruct correlations. The loadings of the individual items on the corresponding variables are all above the recommended value of 0.5, which meets the requirements for appropriate discriminant validity. The loadings range between 0.798 and 0.942. Hence overall the measurement model is highly satisfactory. Table 1 shows a description of the measurement items and their key reliability figures.

Table 1. Measurement item description and reliability analysis

	No. of items	Mean	SD	Cronbach alpha	Composite reliability	AVE
Trust in Internet as booking channel (TR)	3	2.614	.761	.864	.917	.786
Perceived convenience of Internet as booking channel (CONV)	3	2.330	.713	.824	.894	.738
Attitude towards online booking of high-complex travel products (ATT)	3	2.331	.955	.858	.914	.779
Intention to book high-complex travel products online (INT)	3	2.893	1.122	.905	.940	.840
Subjective norm (SN)	4	2.771	.766	.816	.880	.648

5.2 Structural Model

The research model was tested with Partial Least Squares (PLS) analysis using the analysis software SmartPLS [57]. The PLS procedure was applied because according to a Kolmogorov-Smirnoff test the variables are not normal-distributed. Further, the sample size is rather small which makes PLS a more appropriate analysis method. The test of the structural model includes the calculation of the path coefficients, the R-square values of dependent variables, and the t-values. The latter were obtained by a bootstrapping procedure with 100 cases and 1,000 samples and transferred into p values [58].

Table 2 shows the results of the PLS analysis along with the t-values and the p-values of the path coefficients for the research model on online booking intention of high-complex travel products. The R square values of the dependent variables are 0.435 for attitude towards booking high-complex travel products online and 0.338 for intention to book high-complex travel products online.

Table 2. PLS analysis results

Hypothesis	Path coefficient	t-value	p-value	Hypothesis test result
H1: TR -> ATT	.465	3.518	***	Supported
H2: CONV -> ATT	.238	1.878	n.s.	Rejected
H3: ATT -> INT	.453	4.963	***	Supported
H4: SN -> INT	.224	2.380	*	Supported

***p < 0.001; *p < 0.05; n.s. not significant

6 Discussion

The results show that attitude and subjective norm are significant predictors of online booking intention of high-complex travel products. With a path coefficient of 0.453, attitude is the more impactful antecedent of behavioral intention. Thus the results confirm the notion of the theory of planned behavior and numerous findings in the context of e-commerce research (e.g., [49, 59, 60]). Subjective norm turns out to be a significant driver, yet to a lower extent (0.238) which is consistent with some previous e-commerce studies (e.g., [39]) and contrasts with others (e.g., [42]).

Among the independent variables, trust shows a highly significant and very strong impact on attitude towards booking high-complex travel products online (0.465). This result is consistent with previous studies on the role of trust in online purchasing [61, 62] and the role of attitude towards online purchasing as a mediator between trust and online purchase intention [29]. Also research in the tourism context shows a significant impact of trust on online booking intention [63, 64]. Compared with these previous studies that were conducted in the context of low-complex travel products (hotels, airline tickets), the impact of trust is higher for the high-complex travel products.

An opposite finding is achieved on the other independent variable, convenience. In the context of high-complex travel products, convenience turns out to be not significant for attitude formation and online booking intention in the study at hand. This result contrasts with previous studies on the impact of convenience in online booking of travel products. Again, these studies largely investigated low-complex travel products such as hotel reservations [6] or airline tickets [7] or do not address specific travel products [8, 9].

Avenues of explaining the differences between the findings of related studies and those achieved in the study at hand lie in the product complexity. In case of high-complex travel products like circular tours [11, 65], various components are combined that are not only less price-transparent [16], but also less standardized. For instance, individual high-complex travel products may differ by destinations targeted during the tour, hotel selection, group sizes, guides etc. As high-complex travel products also offer a whole package of services, they are more expensive than single travel components of low complexity.

In contrast to low-complex travel products, high-complex travel products are expected to evoke a higher involvement of consumers. In such a setting, perceived product risk [66] may be higher, because consumers may suffer from greater losses if any problems with the product occur. The higher potential risks of damage caused by difficulties in service delivery seem to make consumers more sensitive vis-à-vis the Internet as a channel for booking high-complex travel products online. Moreover, trust appears to become such important that other antecedents become irrelevant. If, for example, a consumer does not sufficiently trust the Internet as a booking channel of high-complex travel products, not even convenience can offset this lack of a necessary prerequisite.

Derived from the noteworthy impact paths of drivers of online booking intention for high-complex travel products, we conclude that the complexity of a travel product can influence the strength of impact of attitude drivers that are known from e-commerce research. More specifically, we propose that complexity of the travel product can moderate the impact of antecedents on attitude towards online booking. A moderating impact of product complexity has already been shown in the context of the impact of product recommendations on search behavior [67].

7 Conclusion

The study at hand empirically tested a research model on drivers of online booking behavior of a high-complex travel product. The findings show that attitude is a strong and subjective norm a medium-strong direct driver of online booking intention which is consistent with the theory of planned behavior and related findings in e-commerce research. When it comes to drivers of attitude, the findings show that the impact of antecedents that are significant in the context of low-complex travel products is different in case of high-complex travel products. Whereas trust in the Internet as a booking channel shows a very high positive impact on attitude, perceived convenience of the Internet as a booking channel has no significant effect. The results therefore suggest that the complexity of the involved product can alter the strength of impact of antecedents that was shown in low-complex travel product contexts.

Like any research, this study has several limitations. One major limitation is the small sample size and the convenience sampling method in the form of a student sample. Whereas this sample is appropriate to gain first insights into the role of travel product complexity, it is neither sufficiently large nor representative to draw any conclusions on a general population. For this purpose, a replication of the study with a larger sample size and the inclusion of further socio-demographic groups (especially in terms of age and education) is necessary. A further limitation is the number of variables included in the research model. For the sake of parsimony, further established variables (e.g., stemming from the technology acceptance model) were not included, but could have a significant or varying impact across travel products at different complexity levels. Further, the study included only organized circular tours as an example of a high-complex travel product. For stronger evidence on the potentially moderating impact of product complexity, further travel products on varying degrees of complexity should be included (e.g., low, medium, high). Finally, the study only considers online channels without differentiating between stationary and mobile devices which gained particular relevance recently [4]. Further research should thus consider more booking channels and include mobile access devices.

References

1. Chiappa, G.D.: Internet versus travel agencies: the perception of different groups of italian online buyers. J. Vacat. Mark. **19**, 55–66 (2013)
2. Gössling, S., Lane, B.: Rural tourism and the development of internet-based accommodation booking platforms: a study in the advantages, dangers and implications of innovation. J. Sustain. Tourism **23**, 1386–1403 (2015)

3. Oppewal, H., Tojib, D.R., Louvieris, P.: Experimental analysis of consumer channel-mix use. J. Bus. Res. **66**, 2226–2233 (2013)
4. Kayak. https://www.kayak.de/news/wp-content/uploads/2016/05/Mobile-Travel-Report-2016. pdf
5. Kayak. https://www.kayak.co.uk/news/wp-content/uploads/sites/5/2016/05/Mobile-Travel-Re port-2016-UK.pdf
6. Izquierdo-Yusta, A., Martinez-Ruiz, M.P., Alvarez-Herranz, A.: What differentiates internet shoppers from internet surfers? Serv. Ind. J. **34**, 530–549 (2014)
7. Yen, H.R.: An attribute-based model of quality satisfaction for internet self-service technology. Serv. Ind. J. **25**, 641–659 (2005)
8. Wen, I.: An empirical study of an online travel purchase intention model. J. Travel Tourism Mark. **29**, 18–39 (2012)
9. Wen, I.: Online shopping of travel products: a study of influence of each dimension of travelers' attitudes and the impact of travelers' online shopping experiences on their purchase intentions. Int. J. Hosp. Tourism Adm. **14**, 203–232 (2013)
10. Anckar, B., Walden, P.: Self-booking of high- and low-complexity travel products: exploratory findings. Inf. Technol. Tourism **4**, 151–165 (2002)
11. Beldona, S., Morrison, A., O'Leary, J.: Online shopping motivations and pleasure travel products: a correspondence analysis. Tourism Manag. **26**, 561–570 (2005)
12. Järveläinen, J.: Online purchase intentions: an empirical testing of a multiple-theory model. J. Organ. Comput. Electron. Commer. **17**, 53–74 (2007)
13. Riley, F.D.O., Scarpi, D., Manaresi, A.: Purchasing services online: a two-country generalization of possible influences. J. Serv. Mark. **23**, 93–103 (2009)
14. Ajzen, I.: The theory of planned behavior. Organ. Behav. Hum. Decis. Process. **50**, 179–211 (1991)
15. Carroll, W.J., Kwortnick, R.J., Rose, N.L.: Travel packaging: an internet frontier. Cornell Hosp. Rep. **17**, 4–16 (2007)
16. Tanford, S., Baloglu, S., Erdem, M.: Travel packaging on the internet: the impact of pricing information and perceived value on consumer choice. J. Travel Res. **51**, 68–80 (2012)
17. Werthner, H., Ricci, F.: E-commerce and tourism. Commun. ACM **47**, 101–105 (2004)
18. Law, R.: Disintermediation of hotel reservations: the perception of different groups of online buyers in Hong Kong. Int. J. Contemp. Hosp. Manag. **21**, 766–772 (2009)
19. Amaro, S., Duarte, P.: Online travel purchasing: a literature review. J. Travel Tourism Mark. **30**, 755–785 (2013)
20. Liu, J.N.K., Zhang, E.Y.: An investigation of factors affecting customer selection of online hotel booking channels. Int. J. Hosp. Manag. **39**, 71–83 (2014)
21. Kim, D.J., Kim, W.G., Han, J.S.: A perceptual mapping of online travel agencies and preference attributes. Tourism Manag. **28**, 591–603 (2007)
22. Wolfe, K., Hsu, C.H., Kang, S.K.: Buyer characteristics among users of various travel intermediaries. J. Travel Tourism Mark. **17**, 51–62 (2005)
23. Ng, E., Cassidy, F., Brown, L.: Exploring the major factors influencing consumer selection of travel agencies in a regional setting. J. Hosp. Tourism Manag. **13**, 75–84 (2006)
24. Chiam, M., Soutar, G., Yeo, A.: Online and off-line travel packages preferences: a conjoint analysis. Int. J. Tourism Res. **11**, 31–40 (2009)
25. Grønflaten, Ø.: Predicting travelers' choice of information sources and information channels. J. Travel Res. **48**, 230–244 (2009)
26. Masiero, L., Law, R.: Comparing reservation channels for hotel rooms: a behavioral perspective. J. Travel Tourism Mark. **33**, 1–13 (2016)
27. Jun, S.H., Vogt, C.A., MacKay, K.J.: Online information search strategies: a focus on flights and accommodations. J. Travel Tourism Mark. **27**, 579–595 (2010)

28. Kim, H.-W., Xu, Y., Gupta, S.: Which is more important in internet shopping, perceived price or trust? Electron. Commer. Res. Appl. **11**, 241–252 (2012)
29. Lim, K.H., Sia, C.L., Lee, M.K.O., Benbasat, I.: Do I trust you online, and if so, will I buy? An empirical study of two trust-building strategies. J. Manag. Inf. Syst. **23**, 233–266 (2006)
30. Pappas, N.: Marketing strategies, perceived risks, and consumer trust in online buying behaviour. J. Retail. Consum. Serv. **29**, 92–103 (2016)
31. Kim, M.-J., Chung, N., Lee, C.-K.: The effect of perceived trust on electronic commerce: shopping online for tourism products and services in South Korea. Tourism Manag. **32**, 256–265 (2011)
32. Kim, M.-J., Lee, C.-K., Chung, N.: Investigating the role of trust and gender in online tourism shopping in South Korea. J. Hosp. Tourism Res. **37**, 377–401 (2013)
33. Christodoulides, G., Michaelidou, N.: Shopping motives as antecedents of e-satisfaction and e-loyalty. J. Mark. Manag. **27**, 181–197 (2011)
34. Hammerschmidt, M., Falk, T., Weijters, B.: Channels in the mirror: an alignable model for assessing customer satisfaction in concurrent channel systems. J. Serv. Res. **19**, 88–101 (2016)
35. Jiang, L.A., Yang, Z., Jun, M.: Measuring consumer perceptions of online shopping convenience. J. Serv. Manag. **24**, 191–214 (2013)
36. Seiders, K., Voss, G.B., Godfrey, A.L., Grewal, D.: Servcon: development and validation of a multidimensional service convenience scale. J. Acad. Mark. Sci. **35**, 144–156 (2007)
37. Nusair, K., Parsa, H.G.: Introducing flow theory to explain the interactive online shopping experience in a travel context. Int. J. Hosp. Tourism Adm. **12**, 1–20 (2011)
38. Chiang, C.-F., Jang, S.C.: The effects of perceived price and brand image on value and purchase intention: leisure travelers' attitudes toward online hotel booking. J. Hosp. Leisure Mark. **15**, 49–69 (2006)
39. Bui, M., Kemp, E.: E-tail emotion regulation: examining online hedonic product purchases. Int. J. Retail Distrib. Manag. **41**, 155–170 (2013)
40. Lee, S., Choi, J., Lee, S.-G.: The impact of a third-party assurance seal in customer purchasing intention. J. Internet Commer. **3**, 33–51 (2004)
41. Njite, D., Parsa, H.G.: Structural equation modeling of factors that influence consumer internet purchase intention of services. J. Serv. Res. **5**, 43–59 (2005)
42. Dennis, C., Jayawardhena, C., Papamatthaiou, E.-K.: Antecedents of internet shopping intentions and the moderating effects of substitutability. Int. Rev. Retail Distrib. Consum. Res. **20**, 411–430 (2010)
43. Besbes, A., Legohérel, P., Kucukusta, D., Law, R.: A cross-cultural validation of the tourism web acceptance model (T-WAM) in different cultural contexts. J. Int. Consum. Mark. **28**, 211–226 (2016)
44. Mouakket, S., Al-Hawari, M.A.: Examining the antecedents of e-loyalty intention in an online reservation environment. J. High Technol. Manag. Res. **23**, 46–57 (2012)
45. Shim, S., Eastlick, M.A., Lotz, S.L., Warrington, P.: An online prepurchase intentions model: the role of intention to search. J. Retail. **77**, 397–416 (2001)
46. Chiou, J.-S.: The antecedents of consumers' loyalty toward internet service providers. Inf. Manag. **41**, 685–695 (2004)
47. Smith, J.B.: Selling alliances: issues and insights. Ind. Mark. Manag. **26**, 146–161 (1997)
48. Taylor, S., Todd, P.A.: Understanding information technology usage: a test of competing models. Inf. Syst. Res. **6**, 144–176 (1995)
49. van der Heijden, H., Verhagen, T., Creemers, M.: Understanding online purchase intentions: contributions from technology and trust perspectives. Eur. J. Inf. Syst. **12**, 41–48 (2003)
50. Gordon, M.E., Slade, L.A., Schmitt, N.: Student guinea pigs: porcine predictors and particularistic phenomena. Acad. Manag. Rev. **12**, 160–163 (1987)

51. Eurostat. http://ec.europa.eu/eurostat/statistics-explained/index.php/Information_society
52. Morris, J.B.: http://www.ntia.doc.gov/blog/2016/first-look-internet-use-2015
53. ITB. http://www.itb-berlin.de/media/itb/itb_dl_all/itb_presse_all/WTTR_Report_2014_Web.pdf
54. Locke, E.A.: Generalizing from laboratory to field: ecological validity or abstraction of essential elements. In: Locke, E.A. (ed.) Generalizing from Laboratory to Field Settings, pp. 3–9. Lexington Books (1986)
55. Nunnally, J.C.: Psychometric Theory. McGraw-Hill, New York (1978)
56. Fornell, C., Larcker, D.F.: Evaluating structural equation models with unobservable variables and measurement error. J. Mark. Res. **18**, 39–50 (1981)
57. Ringle, C.M., Wende, S., Will, A.: www.smartpls.de
58. Rai, A., Patnayakuni, R., Seth, N.: Firm performance impacts of digitally enabled supply chain integration capabilities. MIS Q. **30**, 225–246 (2006)
59. George, J.F.: The theory of planned behavior and internet purchasing. Internet Res. **14**, 198–212 (2004)
60. Pavlou, P.A., Fygenson, M.: Understanding and prediction of electronic commerce adoption: an extension of the theory of planned behavior. MIS Q. **30**, 115–143 (2006)
61. Gefen, D., Karahanna, E., Straub, D.W.: Trust and TAM in online shopping: an integrated model. MIS Q. **27**, 51–90 (2003)
62. Ling, K.C., Chai, L.T., Piew, T.H.: The effects of shopping orientations, online trust and prior online purchase experience toward customers' online purchase intention. Int. Bus. Res. **3**, 63–76 (2010)
63. Gefen, D., Straub, D.W.: Managing user trust in B2C e-services. E-Serv. J. **2**, 7–24 (2003)
64. Nunkoo, R., Ramkissoon, H.: Travelers' e-purchase intent of tourism products and services. J. Hosp. Mark. Manag. **22**, 505–529 (2013)
65. Li, N., Zhang, P.: Consumer online shopping attitudes and behavior: an assessment of research. In: Proceedings of the Eighth Americas Conference on Information Systems, pp. 508–517 (2008)
66. Sorce, P., Perotti, V., Widrick, S.: Attitude and age differences in online buying. Int. J. Retail Distrib. Manag. **33**, 122–132 (2005)
67. Swaminathan, V.: The impact of recommendation agents on consumer evaluation and choice: the moderating role of category risk, product complexity, and consumer knowledge. J. Consum. Psychol. **13**, 93–101 (2003)

Sponsored Data: Smarter Data Pricing in the Age of Data Cap

Xiaowei Mei[(⊠)], Hsing Kenneth Cheng, Subhajyoti Bandyopadhyay, and Liangfei Qiu

Department of Information Systems and Operations Management,
Warrington College of Business Administration, University of Florida,
Gainesville, FL 32611, USA
{xmei,hkcheng,shubho}@ufl.edu,
liangfei.qiu@warrington.ufl.edu

Abstract. As the amount of online content explodes, consumers have to make a choice: either cut down their mobile data consumption or pay high overage fees. We investigate a recent phenomenon whereby network service providers are encouraging content providers to sponsor data for consumers. We analyze this phenomenon using game theory within a setting of one monopoly mobile network operator (MNO) and two competing content providers (CPs). Consumers are heterogeneous in both data usage and in their preference for the CPs. We find that the MNO's optimal profit decreases as the consumers' data cap becomes larger, and the optimal pricing scheme is a two-part tariff without any data caps.

Keywords: Data cap · Zero rating · Sponsored data · Two-part tariff

1 Introduction

With the development of data-intensive internet services, consumers' mobile data quotas can easily be exhausted by, for example, watching a few HD videos. According to Cisco's Visual Network Indexing, mobile data traffic will grow 10-fold from 2014 to 2019 at a compound annual growth rate of 57%, which is three times faster than broadband network. By 2019, there will be 5.2 billion global mobile users and 11.5 billion mobile-ready devices and connections (Cisco 2015). The whole world is going mobile.

The mobile network operators (MNOs), such as Verizon, AT&T, and T-Mobile, would like this trend to continue: increased data consumption begets newer types of content from innovative content providers (CPs), such as Netflix and Spotify, which in turn encourages consumers to consume more. Consumers, meanwhile, are becoming more conscious about their data consumption, as their monthly cellphone charges are becoming increasingly large. In view of this, the MNOs have recently proposed a plan to transfer at least a fraction of the data bill from the consumers to the CPs. In their 2014 Developer Summit, executives of AT&T introduced "Sponsored Data", also called 1-800-DATA, which allows eligible 4G customers to enjoy more mobile content and apps over AT&T's wireless network without having their consumption counted

© Springer International Publishing AG 2017
M. Fan et al. (Eds.): WeB 2016, LNBIP 296, pp. 150–161, 2017.
https://doi.org/10.1007/978-3-319-69644-7_15

towards their monthly data plans. One year after their proposal, 10 companies have signed up and AT&T is looking for more.[1] Other examples include Amazon which pays AT&T for the wireless connectivity that allows it to deliver electronic books to people free of connectivity charges to its Kindle readers.[2] ESPN has had discussions with at least one major U.S. carrier to subsidize wireless connectivity on behalf of its users.[3] Verizon too has confirmed to join AT&T in charging companies for sponsored data at the end of 2015.[4] Customers of T-Mobile started enjoying free music from top streaming services, including Spotify, Pandora, iTunes Radio, etc. since 2014 (Gryta 2014). In November 2015, they launched a new service called Binge On that offers consumers free streaming video from forty-two providers – Netflix, Amazon, Hulu, HBO, among others – without using their data plans (Gryta and Knutson 2015). Although sponsored data is a relative newcomer to the U.S. cellular market, it has actually been used in other venues and regions for some time in the form called "zero rating". Since 2010, Facebook has worked with more than 50 mobile operators in 45 countries and territories to provide a stripped-down text-only version of its mobile website with zero data charges.[5] Later in 2012, eBay joined a growing list of top shopping and e-commerce sites that can be found on the inflight internet service provider Gogo's multimedia platform, which offers passengers of airlines such as Delta and Virgin America free access to eBay.[6] This partnership with Gogo allows customers of eBay to shop more than 300 million listings on the website without worrying about data usage while traveling above 10,000 feet. Google gives users in select countries the freedom to use Google Search, check Gmail and socialize with Google Plus without incurring any data fees through Google Free Zone since 2013.

While being quickly applauded by many consumers, such a business model has been criticized by advocators of net neutrality (Cheng et al. 2011, Economides and Hermalin 2012), who contend that it puts small businesses and developers at a distinct disadvantage to their deeper-pocketed counterparts who can pay to exempt their traffic from consumers' bills in a tiered mobile internet. Such a development, they feel, would hurt innovation in mobile internet in the long run. They are also concerned that Internet service providers (ISP) could also use its position as a gatekeeper to pick winners and losers online, which would distort competition. Telecom companies, however, argue that this subsidization model does not violate the principle of net neutrality since sponsored data is transmitted without priority over non-sponsored data. Others claim that AT&T's double dipping, or two-sided billing, actually places more burden on content providers by compelling them to incur an additional cost just to get their

---

[1] http://arstechnica.com/business/2015/01/att-has-10-businesses-paying-for-data-cap-exemptions-and-wants-more/.

[2] https://gigaom.com/2012/07/24/amazon-limits-monthly-kindle-browsing-over-3g-to-50-mb/.

[3] http://www.wsj.com/articles/SB10001424127887324059704578473400083982568.

[4] https://www.washingtonpost.com/news/the-switch/wp/2015/12/10/verizon-to-start-offering-sponsored-data/.

[5] http://techcrunch.com/2010/05/18/facebook-launches-0-facebook-com-a-mobile-site-that-incurs-zero-data-fees/.

[6] http://thenextweb.com/us/2012/08/13/in-flight-wifi-company-gogo-now-offers-delta-virgin-america-customers-free-access-ebay/#gref.

content delivered to customers.[7] There are yet others who are concerned that satisfied with "free" access to a walled garden of chosen services, people who rely on sponsored data may never move to "real" Internet access.[8] India just blocked Facebook's Free Basics internet which provides free content to people who don't have access to internet. This included selected local news, weather forecasts, the BBC and some health sites. However, the Telecom Regulatory Authority of India rules that "no service provider shall offer or charge discriminatory tariffs for data services on the basis of content".[9]

In this paper, we analyze this new pricing model in cellular networks within a game theoretical framework. We derive the incentive and solutions for the MNO to optimize its profit under different scenarios where there are multiple CPs who can choose whether or not to participate in sponsored data. We find that the MNO's optimal profit is lesser when consumers have a data cap and the optimal pricing scheme is a two-part tariff without any data cap, in other words, a usage-based pricing plus a fixed access fee. This pricing structure has been adopted by some industry pioneers. Google Fi, which is a mobile virtual network operator (MVNO) owned by Alphabet, charges users $20/month for connectivity and $10/GB of data consumed (Womack and Moritz 2015). At the end of the month it refunds consumers for any data purchased but not used. Republic Wireless, another MVNO, also announced a similar offer (Goldstein 2015).

A unique feature of our model is that we allow both individual usage heterogeneity and horizontal preference differentiation by using a two-dimensional Hotelling model. Consumer heterogeneity on both dimensions plays a significant role in the profits that the MNO and the CPs make, which directly affects the strategy they choose for sponsoring data. Our model allows us to capture this market reality.

2 The Model

A monopolistic mobile network operator (MNO) provides mobile network access to consumers. In order to get access to the network, consumers have to pay a fixed up-front fee F which is the same for all customers. There's a data cap Λ under which customers can consume as much wireless data as they want without being charged any additional fee. However, if the amount of data they consume exceeds the data cap, over-used data will be charged at price p per unit of data. There are two content providers (CPs), L and H, who provide online content through the MNO's network to consumers for free and generate revenues from advertisers who want to reach these consumers. Let r_L and r_H denote the average revenue per unit (ARPU) of L and H, respectively. Without loss of generality, we assume $r_L < r_H$, which means that content provider H is better than L in generating profit. Consumers choose their preferred CP to consume the online content.

---

[7] http://www.forbes.com/sites/ewanspence/2012/02/27/att-looking-to-double-dip-on-mobile-data-char ges/#1800acc372f6.

[8] http://www.theverge.com/2014/1/6/5280566/att-sponsored-data-bad-for-the-internet-the-economy-and-you.

[9] http://www.thehindu.com/sci-tech/technology/internet/trai-rules-in-favour-of-net-neutrality/article82 09455.ece.

In order to attract more consumers and increase market share, content providers can participate in sponsored data plans by paying the overage bill for consumers. Depending on whether content provider L and/or H participate in sponsored data plans, there are four cases: Case 1 – neither content provider sponsors data, Case 2 – only content provider L sponsors data, Case 3 – only content provider H sponsors data, Case 4 – both content providers sponsor data. We discuss each case in the following subsections. Table 1 summarizes all the cases with π_L and π_H denoting the profit of L and H respectively and the superscripts referring to the various subcases that we encounter depending on the prevailing market conditions.

Table 1. Summary of four cases

Strategy of CP		CP H	
		Not sponsoring	Sponsoring
CP L	Not sponsoring	Case 1: π_L^1, π_H^1	Case 3A: π_L^{3A}, π_H^{3A}
			Case 3B: π_L^{3B}, π_H^{3B}
	Sponsoring	Case 2A: π_L^{2A}, π_H^{2A}	Case 4A: π_L^{4A}, π_H^{4A}
		Case 2B: π_L^{2B}, π_H^{2B}	Case 4B: π_L^{4B}, π_H^{4B}

In our model, consumers are heterogeneous in two dimensions, individual preference for content provider (denoted by x) and amount of data consumption (denoted by k). Let the consumers' ideal content provider be uniformly distributed on a horizontal line segment [0, 1] with content provider L located at 0 and H at 1. Denote $\underline{k}$ to be the lowest data consumption of consumers and $\bar{k}$ the highest. Without loss of generality, we normalize $[\underline{k}, \bar{k}]$ to [0, 1]. Thus each consumer can be represented with a tuple (x, k), and they are uniformly distributed in a unit square. We assume that the MNO has full market coverage, so every consumer in the square consumes wireless data from MNO.

The MNO's profit composes of two parts: fixed fee F from all consumers and overage fee charged for data consumption exceeding Λ, either from the consumers when there is no sponsored data or from the content providers when there's sponsored data. Therefore MNO's profit is

$$\pi_{MNO} = \int_0^1 \int_0^1 \left[F + (k - \Lambda)^+ p \right] dk dx = F + \frac{p}{2}(1 - \Lambda)^2 \tag{1}$$

2.1   Case 1: Neither CP Provides Sponsored Data

When no content provider participates in sponsored data, consumers have to pay for data usage over the cap by themselves. A consumer's net utility from consuming online content by either content provider L or H depends on one's individual preference, the distance of one's preferred provider (i.e., either L or H) from one's ideal, and the amount of data one consumes. Let parameter t measure the unit transportation cost of the deviation from a consumer's ideal content in the Hotelling framework (Hotelling 1929). Following Mendelson (1985), we denote $V(k)$ to be the gross value function of a consumer when k unit of data is consumed. $V(k)$ is twice differentiable and strictly concave. Each consumer is charged a fixed up-front fee F to get access to the network. Therefore the utility function for an arbitrary customer (x, k) who chooses content provider L is

$$U_L^1(x, k) = V(k) - \left[F + (k - \Lambda)^+ p\right] - xt \tag{2}$$

and

$$U_H^1(x, k) = V(k) - \left[F + (k - \Lambda)^+ p\right] - (1 - x)t \tag{3}$$

if content provider H is chosen. Setting these two utility functions equal, we get the marginal consumer line $x^* = \frac{1}{2}$. Consumers on the left of this line choose L, while consumers one the right choose content provider H. Market share is allocated equally between L and H and is shown in Fig. 1.

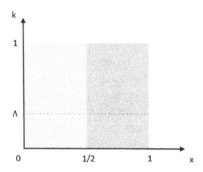

Fig. 1. Market share under Case 1

Revenues of the CPs come from advertisers paying for the consumers' click-throughs and we assume that all other costs for the CPs are sunk cost and are normalized to zero. Thus profits for content provider L and H are

$$\pi_L^1 = \int_0^{\frac{1}{2}} \int_0^1 (r_L k) dk dx = \frac{1}{4} r_L \tag{4}$$

$$\pi_H^1 = \int_{\frac{1}{2}}^1 \int_0^1 (r_H k) dk dx = \frac{1}{4} r_H \tag{5}$$

In order to cover the entire market, the MNO has to make the utility of every consumer nonnegative. The MNO's profit-maximization problem therefore is

$$\max_{F,p} F + \tfrac{p}{2}(1 - \Lambda)^2$$

$$s.t. U_L^1(x, k) \geq 0 \text{ for } 0 \leq x \leq \tfrac{1}{2}, 0 \leq k \leq 1,$$

$$U_H^1(x, k) \geq 0 \text{ for } \tfrac{1}{2} < x \leq 1, 0 \leq k \leq 1, \qquad (6)$$

$$F \geq 0, p \geq 0.$$

The first two constraints are participation constraints for consumers of L and H, respectively. These constraints ensure all the consumers' utilities to be nonnegative. Solving the MNO's problem, we have the following result.

Proposition 1 (MNO's Pricing Strategy): *When neither content provider sponsors data, the overage price charged by MNO increases in the data cap. MNO's profit decreases in both the unit transportation cost and the data cap.*

Proof. Available upon request.

The solution of Proposition 1 shows that the fixed up-front fee is determined by the consumer with the lowest utility in the market, located at $\left(\tfrac{1}{2}, 0\right)$. This consumer incurs the highest transportation cost and consumes the least amount of data and is indifferent between joining the network and not joining it. High unit transportation cost hurts the utility of this consumer. To ensure the participation of this consumer, the up-front fee decreases as the unit transportation costs get higher, which in turn lowers the MNO's profit. If the MNO sets the data cap at a high level, the consumers would be allowed to consume a large amount of data without being charged an overage fee. To maintain the same level of profitability, the MNO would set a high price for data usage above the cap to more effectively extract the surplus from the heavy users. However, our analysis actually suggests that in order to maximize profit, the MNO should set the data cap as low as possible. In other words, a pure usage pricing without any data cap would maximize the MNO's profit under this case. The intuition is quite straightforward – charging users for every bit they consume allows the MNO to perfectly discriminate between all the consumers. Therefore, there won't be "free" data (i.e. under the cap) anymore.

2.2 Case 2 and Case 3: Only One CP Provides Sponsored Data

When one of the CPs provides sponsored data, consumers who choose this CP will not pay for the overage fee if their data consumption exceeds the cap. The MNO simply transfers the bill for overage data to this CP. Thus, some consumers who used to consume content from the other CP would switch to this one in order to get free data although the other CP might fit them better. In this section we illustrate the case when only CP L provides sponsored data as Case 2. The case when only CP H sponsors data, denoted as Case 3, is interchangeable.

The utility function for an arbitrary consumer (x, k) becomes

$$U_L^2(x, k) = V(k) - F - xt \qquad (7)$$

if she consumes content from CP L, and

$$U_H^2(x, k) = V(k) - \left[F + (k - \varLambda)^+ p\right] - (1 - x)t \tag{8}$$

if she chooses CP H. So when $0 \le k \le \varLambda$, the marginal line is $x = \frac{1}{2}$; when $\varLambda < k \le 1$, the marginal line is $k = \frac{2t}{p}x + \varLambda - \frac{t}{p}$. L captures more market share than H since the heavy users of H cannot afford the high overage price and are willing to bear a little more disutility (in terms of fit) to get the free data after switching over to L. As the overage price p charged by the MNO increases, the slope of the marginal line $\frac{2t}{p}$ declines. Depending on whether $p \le \frac{t}{1-\varLambda}$ or $p > \frac{t}{1-\varLambda}$, there are two subcases when only content provider L sponsors data.

2.2.1   Case 2A: $p \le \frac{t}{1-\varLambda}$
When the overage price charged by the MNO is not too high, the market share of L and H can be shown as in Fig. 2.

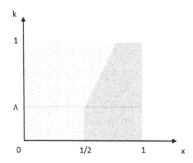

Fig. 2. Market share under Case 2A

Revenue for content provider L comes from consumers' click-throughs, however, L also bears the cost of sponsoring data. The profit of L thus is

$$\pi_L^{2A} = \int_0^{\frac{1}{2}} \int_0^{\varLambda} (r_L k)dkdx + \int_0^{\frac{1}{2}} \int_{\varLambda}^{1} (r_L - p)kdkdx + \int_{\varLambda}^{1} \int_{\frac{1}{2}}^{\frac{pk}{2t} - \frac{p\varLambda}{2t} + \frac{1}{2}} (r_L - p)kdxdk \tag{9}$$

Since H does not sponsor data, the profit for H is

$$\pi_H^{2A} = \int_{\frac{1}{2}}^{1} \int_0^{\varLambda} (r_H k)dkdx + \int_{\varLambda}^{1} \int_{\frac{pk}{2t} - \frac{\varLambda p}{2t} + \frac{1}{2}}^{1} (r_H k)dxdk \tag{10}$$

Compared to Case 1, the profit of H decreases because part of its market share is captured by L. However, H still makes a nonnegative profit. To make L sponsor data, the MNO has to guarantee that L makes no less profit than when it does not sponsor data. Thus, the MNO's problem becomes:

$$\max_{F,p} F + \frac{p}{2}(1 - \Lambda)^2$$

s.t. $U_L^2(x,k) \geq 0$ for $0 \leq x \leq \frac{1}{2}, 0 \leq k \leq \Lambda$, or $0 \leq x \leq \frac{pk}{2t} - \frac{p\Lambda}{2t} + \frac{1}{2}, \Lambda \leq k \leq 1$,

$U_H^2(x,k) \geq 0$ for $\frac{1}{2} < x \leq 1, 0 \leq k \leq \Lambda$ or $\frac{pk}{2t} - \frac{p\Lambda}{2t} + \frac{1}{2} \leq x \leq 1, \Lambda \leq k \leq 1$,   (11)

$$\pi_L^{2A} \geq \frac{1}{4} r_L,$$

$$F \geq 0,$$

$$0 \leq p \leq \frac{t}{1-\Lambda}.$$

The first two constraints are participation constraints for consumers of L and H, respectively. The third constraint is the participation constraint for L to sponsor data.

2.2.2 Case 2B: $p > \frac{t}{1-\Lambda}$

When the overage price charged by the MNO gets higher, L captures more market share of its competitor and the market shares of L and H can be shown in Fig. 3.

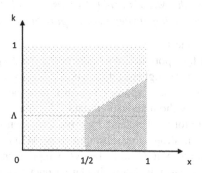

Fig. 3. Market share under Case 2B

Profit for content provider L becomes

$$\pi_L^{2B} = \int_0^{\frac{1}{2}} \int_0^1 (r_L k) dk dx + \int_{\frac{1}{2}}^1 \int_{\Lambda - \frac{t}{p} + \left(\frac{2t}{p}\right)x}^1 (r_L k) dk dx$$
$$- \int_0^{\frac{1}{2}} \int_\Lambda^1 (pk) dk dx - \int_{\frac{1}{2}}^1 \int_{\Lambda - \frac{t}{p} + \left(\frac{2t}{p}\right)x}^1 (pk) dk dx$$ (12)

The profit of H is given by

$$\pi_H^{2B} = \int_{\frac{1}{2}}^1 \int_0^{\Lambda - \frac{t}{p} + \left(\frac{2t}{p}\right)x} (r_H k) dk dx$$ (13)

The MNO's profit-maximization problem is as follows:

$$\max_{F,p} F + \tfrac{p}{2}(1-\varLambda)^2$$

$$s.t. \ U_L^2(x,k) \geq 0 \text{ for } 0 \leq x \leq \tfrac{1}{2}, 0 \leq k \leq 1, \text{ or } \tfrac{1}{2} \leq x \leq 1, \varLambda - \tfrac{t}{p} + \left(\tfrac{2t}{p}\right)x \leq k \leq 1,$$

$$U_H^2(x,k) \geq 0 \text{ for } \tfrac{1}{2} < x \leq 1, 0 \leq k \leq \varLambda - \tfrac{t}{p} + \left(\tfrac{2t}{p}\right)x,$$

$$\pi_L^{2B} \geq \tfrac{1}{4}r_L,$$

$$F \geq 0,$$

$$p > \tfrac{t}{1-\varLambda}.$$

$$(14)$$

Again, the first two constraints are participation constraints for consumers of L and H, respectively. The third constraint is the participation constraint for L to sponsor data. Solving the MNO's problem, we have the following result.

Proposition 2 (MNO's Strategy): *When only L sponsors data, the overage price charged by the MNO and the MNO's profit decrease in unit transportation cost and data cap. MNO's optimal pricing strategy is two-part tariff without data cap.*

Proof. Available upon request.

The consumers' unit transportation cost affects the fixed up-front fee in the same way as in Case 1. Furthermore, higher unit transportation cost implies higher switching costs for consumers, which makes sponsored data less attractive for those loyal consumers who prefer the CP without sponsored data. This makes it more difficult for the sponsoring CP to compete for market share by providing sponsored data and leads to a lower overage price he can afford, thus hurts profit of MNO who prefers the highest feasible overage price. Contrary to Case 1, the overage price charged by MNO actually decreases in data cap. The reason is that with a larger data cap, the market share left for CPs to compete by sponsoring data is diminished. This makes it less attractive for CPs to provide sponsored data and the affordable overage price for CPs also drops, which in turn lowers the surplus that the MNO can extract.

Under Case 3, H provides sponsored data instead of L. This Case is symmetric to Case 2 except that H has a higher ARPU, and therefore has the incentive to pay for a higher overage price charged by MNO. Under this case, we have similar results for the MNO's optimal strategy, i.e. a two-part tariff without a data cap is optimal.

2.3   Case 4: Both CPs Provide Sponsored Data

When both content providers sponsor data, consumers will not have to pay the overage fee if the amount of data they consume exceeds the data cap no matter which CP they choose. Therefore, the utility function for an arbitrary customer (x, k) is

$$U_L^4(x,k) = V(k) - F - xt \tag{15}$$

if she chooses L and is

$$U_H^4(x, k) = V(k) - F - (1 - x)t \tag{16}$$

if she chooses H. The market share of content provider L and H is the same as in Case 1. Each consumer will stick with their preferred content provider and the marginal consumers still locate on the line $x^* = \frac{1}{2}$. The following figure shows market share of content provider L and H under Case 4 (Fig. 4).

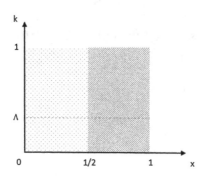

Fig. 4. Market share under Case 4

The profit for content provider L is

$$\pi_L^4 = \int_0^{\frac{1}{2}} \int_0^1 (r_L - p)k\,dk\,dx = \frac{1}{4}(r_L - p) \tag{17}$$

And the profit for content provider H is

$$\pi_H^4 = \int_{\frac{1}{2}}^1 \int_0^1 (r_H - p)k\,dk\,dx = \frac{1}{4}(r_H - p) \tag{18}$$

2.3.1 Case 4A: $p \le \frac{t}{1-\Lambda}$

The MNO's profit-maximization problem is as follows:

$$\max_{F,p} F + \frac{p}{2}(1 - \Lambda)^2$$

$$s.t.\ U\left(\tfrac{1}{2}, 0\right) = V(0) - F - \tfrac{1}{2}t \ge 0,$$

$$\tfrac{1}{4}(r_L - p) \ge \tfrac{1}{4}r_L(2 - \Lambda^2) - \tfrac{r_L r_H}{12t}(\Lambda^3 - 3\Lambda + 2),$$

$$\tfrac{1}{4}(r_H - p) \ge \tfrac{1}{4}r_H(2 - \Lambda^2) - \tfrac{r_L r_H}{12t}(\Lambda^3 - 3\Lambda + 2),$$

$$F \ge 0,$$

$$0 \le p \le \tfrac{t}{1-\Lambda}. \tag{19}$$

The first constraint is consumers' participation constraint. Similar to Case 1, the consumer with lowest utility is located at $\left(\frac{1}{2}, 0\right)$ since she consumes the least amount of data but incurs highest transportation cost. The second and third constraints are the participation constraints for content provider L and H, respectively.

2.3.2 Case 4B: $p > \frac{t}{1-\Lambda}$

The MNO's profit-maximization problem is as follows:

$$\max_{F,p} F + \frac{p}{2}(1 - \Lambda)^2$$

$$s.t.\ U\left(\tfrac{1}{2}, 0\right) = V(0) - F - \tfrac{1}{2}t \geq 0,$$

$$\tfrac{1}{4}(r_L - p) \geq \tfrac{1}{4}r_L - \frac{r_L(1-\Lambda^2)p_{3B}^*}{4\left(r_H - p_{3B}^*\right)},$$

$$\tfrac{1}{4}(r_H - p) \geq \tfrac{1}{4}r_H - \frac{r_H(1-\Lambda^2)p_{2B}^*}{4\left(r_L - p_{2B}^*\right)}, \tag{20}$$

$$F \geq 0,$$

$$p > \tfrac{t}{1-\Lambda}.$$

Proposition 3 (MNO's Strategy): *When both CPs sponsor data, the overage price charged by the MNO and the MNO's profit decrease in the unit transportation cost and data cap. The MNO's optimal pricing strategy is a two-part tariff without data cap.*

Proof. Available upon request.

3   Summary and Conclusion

We analyze the pricing scheme for a monopoly mobile network operator and two content providers who might subsidize consumers by providing sponsored data. We incorporate consumer heterogeneity in two dimensions, namely data usage and horizontal preference for the two CPs, by applying a two-dimension Hotelling model. We find that the current nonlinear pricing scheme of having a fixed fee for data consumption up to a limit or cap might not be optimal for the MNO. The MNO can actually extract more surplus from consumers by removing the cap. Such a pricing structure has actually been adopted by several industry pioneers (Womack and Moritz 2015, Goldstein 2015). Our finds have managerial implications for network operator practitioners.

References

Womack, B., Moritz, S.: Google Announces Project Fi Wireless Network Starting at $20 per Month. BloombergBusiness (2015). http://www.bloomberg.com/news/articles/2015-04-22/google-is-said-to-be-unveiling-mobile-phone-service-today

Cheng, H.K., Bandyopadhyay, S., Guo, H.: The debate on net neutrality: a policy perspective. Inf. Syst. Res. **22**(1), 60–82 (2011)

Cisco: Cisco Visual Networking Index: Forecast and Methodology. Cisco 2014–2019 (2015). White Paper. http://www.cisco.com/c/en/us/solutions/collateral/service-provider/ip-ngn-ip-next-generation-network/white_paper_c11-481360.html

(GLOBECOM). IEEE 2013

Economides, N., Hermalin, B.: The economics of network neutrality. RAND J. Econ. **43**(4), 602–629 (2012)

Economides, N., Hermalin, B.E.: The strategic use of download limits by a monopoly platform. RAND J. Econ. **46**(2), 297–327 (2015)

Ma, R.T.: Subsidization competition: vitalizing the neutral Internet. In: Proceedings of the 10th ACM International on Conference on emerging Networking Experiments and Technologies, pp. 283–294. ACM (2014)

Nabipay, P., Odlyzko, A., Zhang, Z.-L.: Flat versus metered rates, bundling, and bandwidth hogs. In: 6th Workshop on the Economics of Networks, Systems, and Computation (2011)

Goldstein, P.: Republic Wireless launches plans that refund customers for unused data. FierceWireless (2015). http://www.fiercewireless.com/story/republic-wireless-launches-plans-refund-customers-unused-data/2015-07-07

Gryta, T.: T-mobile will waive data fees for music services. Wall Str. J. (2014). http://www.wsj.com/articles/t-mobile-will-waive-data-fees-for-music-service-1403142678

Gryta, T., Knutson, R.: T-mobile to offer free video streaming. Wall Str. J. (2015). http://www.wsj.com/articles/t-mobile-to-offer-free-video-streaming-1447187527

Please Share! Online Word of Mouth and Charitable Crowdfunding

Mahdi Moqri[(⊠)] and Subhajyoti Bandyopadhyay

University of Florida, Gainesville, FL, USA
moqri@ufl.edu

Abstract. During the last few years, charitable crowdfunding has become an increasingly popular method of online fundraising for personal and charitable causes. Many such crowdfunding platforms encourage and facilitate the use of online word of mouth (WOM) through social networks and social media, to spread the word about the crowdfunding campaigns. While online WOM is commonly used to share information about crowdfunding campaigns, there is hitherto limited understanding as to whether or how this information sharing affect individuals' contribution behavior or the outcome of crowdfunding campaigns. In this study, using a unique dataset from 590 crowdfunding campaigns observed over 12 days, we examine to what extent, and how quickly online WOM affect the rate of contributions. In addition, we explore the effect of different phases of fundraising (over time or as they approach their target goals) and of the coverage of the campaigns in major online news websites.

Keywords: Charitable crowdfunding · Social media · Online WOM · Word of mouth

1 Introduction

Over the past few years, crowdfunding platforms have been used increasingly to raise funds for charitable causes. Although crowdfunding was developed and used initially to fund creative and for-profit projects (e.g., music and video game projects on Kickstarter: Agrawal et al. 2013, Lehner 2013), new platforms, including GoFundMe, CrowdRise, and DonorChoose have been developed primarily to raise funds for individuals with personal needs. Currently, the largest crowdfunding website in the world (based on the number of visitors) is the donation-based website GoFundMe.com which has raised over $1 billion (www.gofundme.com). Four out of ten of the most popular platforms are devoted primarily to personal or donation-based fundraising (based on Alexa & Compete web traffic data reported on www.gofundme.com). Despite the increasing growth and widespread use and popularity of such platforms, there is limited understanding in the literature as to how campaigns raise money on the platforms.

Crowdfunding platforms often facilitate and encourage "sharing the campaign" with others through social networks. For example, GoFundMe recommends Facebook as "the absolute best way to reach out to those closest to you," and claims that sharing

M. Fan et al. (Eds.): WeB 2016, LNBIP 296, pp. 162–169, 2017.
https://doi.org/10.1007/978-3-319-69644-7_16

the campaign on Facebook increases donations by 350%.[1] Nonetheless, up to an estimated 40% of organizers (based on our sample) do not share their campaigns on Facebook. Few studies have explored the relationship between online "sharing" (i.e., online/electronic word of mouth) and the outcome of crowdfunding campaigns especially in the context of personal and charitable crowdfunding. Furthermore, the existing studies do not agree on the effect of various means of WOM.

Using a unique set of panel data from 590 crowdfunding campaigns observed over 48 periods of 6 h each, we examine how online word of mouth (WOM) affects contribution behaviors. Specifically, we estimate to what extent, and how quickly, Facebook shares and Twitter posts affect the rate of contributions. In addition, we explore the effect of different phases of fundraising (over time or as they approach their target goals) and the impact of the coverage of campaigns in major online news websites. Finally, we examine and control for the effect of previous campaign contributions (reinforcement or substitution effect).

2 Theoretical Background

2.1 WOM, Awareness and Social Influence

WOM on social media and social network platforms (hereafter called social platforms) play a role in charitable giving through two different mechanisms: (a) by creating awareness and (b) through social influence. Awareness has been well recognized as the first step in the charitable giving decision process (Guy and Patten 1989; Snipes and Oswald 2010). The popularity of social platforms have enhanced the efficiency and effectiveness of communication, by providing a means to reach a large audience at a low cost. These platforms have enabled individuals to contribute to charitable causes using their social capital. Even those who do not donate directly to the charitable cause can contribute towards it by spreading the word and informing potential donors about it through social platforms.

In addition to awareness, social platforms can create social influence or facilitate donors' social objectives. Social objectives, such as reputation and desire for respect or social acclaim, have been theoretically and experimentally shown to positively affect people's motivation to donate. Harbaugh (1998) and Bénabou and Tirole (2005) theoretically modeled how desire for prestige and social reputation influences donors' behaviors. Saxton and Wang's (2014) study suggested that "social network effect takes precedence over traditional economic explanations" in charitable giving within a social network. Similarly, Castillo et al.'s (2014) experiments demonstrated how public announcement of donations on an online social network affect friends' donations. These public announcements are seen as a soft ask (diffuse ask) to many friends on the social platform, which might motivate donation by creating an indirect social pressure.

We investigate three main channels through which campaigns raise awareness or create social influence: (1) online WOM channels (e.g., Facebook and Twitter), (2) offline WOM channels (e.g., face-to-face conversations), and (3) mass media (e.g., news

---

[1] http://support.gofundme.com/hc/en-us/articles/203604494-6-Steps-to-a-Successful-Campaign.

and broadcasts). We hypothesize that these channels affect the intention to donate positively by increasing the awareness about the campaigns or by creating social influence or by enhancing donors' social objectives.

2.2   WOM and Donation-Based Crowdfunding

Crowdfunding platforms, both for-profit (reward-based) and donation-based, help raise monetary contribution from many individuals in two forms: (1) by facilitating making an open call to the public to raise monetary contributions, and (2) by managing the transfer and collection of the funds. These platforms offer a web-based space for a project/campaign where the organizers describe the investment opportunity or the need. Many platforms also facilitate and encourage communicating the project/campaign through social platforms as well. However, few large-scale studies have empirically demonstrated the effect of sharing on social platforms on the contribution to crowdfunding.

Lu et al.'s (2014) study was the first to report that the number of posts on Twitter (i.e., tweets) about a project on Kickstarter was correlated positively with the outcomes of the project's fundraising efforts. However, they claimed that the correlations were "primarily due to the fact that if a project is more persuasive to intrigue authors in social networks to discuss it (either by social promotions or external stimulations), it is more attractive to investors." Thies et al. (2014) observed that an increase in the total number of daily Facebook posts (shares) about a crowdfunding campaign had a positive effect on the number of backers (funders) on the following day, while they observed either no effect, or a negative effect, for daily tweets. The data used for their study were collected from Indiegogo. The authors proposed that because those who fund successful projects typically receive some form of reward (or return on their investments), financial incentives (e.g., minimizing the uncertainty of backing a project) might influence a funder's sharing and investing behavior. To the best of our knowledge, the only study that has focused on the effect of online WOM in donation-based crowdfunding is that of Hong et al. (2015). These authors found empirical evidence to support the hypotheses that an increase in the number of Facebook shares only had positive effects on "public good" (e.g., charitable) campaigns, while tweets only affected "private good" (e.g., creative product) campaigns positively. The proposed hypotheses were based on the assumption that Facebook users are more responsive to desirable behaviors in a social group, such as donations to charity, while Twitter users are more responsive to consumer goods and services.

2.3   Perception of Urgency

The literature on charitable giving has established a positive effect of the perception of urgency on giving behavior Sargeant (1999). In addition to campaigns' inherent characteristics, the passage of time and funding progress might affect the perception of urgency. We hypothesize that as campaigns become "older" on the platform, or as they approach their target goals, they are perceived to be less urgent. Therefore, both time and funding progress are expected to affect perceived urgency negatively and consequently, reduce the intention to donate.

3 Data

We collected a comprehensive set of information on 590 crowdfunding campaigns from GoFundMe, a donation-based (as compared to reward-based) crowdfunding platform that allows people to raise money for personal causes (e.g., medical expenses, education costs, funerals, and memorials) or charities. Users can create their own campaign webpages to describe why they raise money, how much money they need to raise, and can post updates about the cause. The data were collected from multiple sources. First, each campaign's information, including its description, organizer, target goal, and information about all contributions, was collected every 6 h using a web-crawler application that we developed. Second, a Twitter streaming API was established to collect complete information on each post (and its writer) for each campaign. Third, the number of Facebook shares about each campaign during each period was collected using a Facebook API embedded in the crowdfunding platform. Fourth, a separate web-crawler application was developed to collect information about any news coverage of each campaign by any of the online news websites indexed by Google News, during each period.

The majority of crowdfunding campaigns on GoFundMe raise money for personal causes, such as covering medical costs. The target goals of most campaigns (the total amount of money requested from the crowd) are somewhere between several hundreds and several thousands of dollars. While more than 90% of campaigns in our dataset raised less than $500, a few attracted tens of thousands of dollars.

3.1 Campaign Characteristics

The goals of the campaigns varied significantly in urgency, ranging from non-urgent (e.g. an educational trip) to life threatening conditions (e.g. a critical surgery). Moreover, the campaigns might have been setup by people with varying levels of experience or different numbers of friends or followers on social networks. Such characteristics of a campaign affect its popularity and attractiveness both to funders and other people who visit the campaign's' webpage. To control for the urgency of the causes, fundraisers' experience, and all other time-invariant factors, we collected multiple observations for each campaign and employed a fixed effect (within effect) panel data model.

3.2 Time and Funding Progress

All of the campaigns in our dataset were started on the same day, January 18, 2016: this decreased the effect of seasonal or time-dependent factors, and facilitated examining and controlling for them. In addition, it allowed us to observe these campaigns from their inception. The largest number of contributions is typically made in the first day of a campaign, and thereafter, the number of contributions decrease exponentially over time. Overall, approximately 95% of contributions are made during the first week of a campaign.

In order to visually examine how contribution rates changed as campaigns approached their goal, we first chose all campaigns which reached at least 30% of their goals (78 campaigns). For each of these campaigns, we then plotted the cumulative contribution as a function of time. While we observed varying patterns of progress, the most frequently observed pattern was a "bursty" starting phase followed by a long period of nearly flat cumulative contributions. We did not observe a "bursty" final phase such as the one reported by Lu et al. (2014) for Kickstarter campaigns.

3.3   Online Word of Mouth

We measured online WOM using the two sources: shares on Facebook and tweets. The distribution of both the number of Facebook shares and tweets about campaigns were extremely skewed. Similar to the number of contributions, online WOM factors decreased sharply after the first few days. The total number of contributions, Facebook shares and tweets were correlated strongly and positively. These correlations might be caused by campaigns' characteristics (e.g. urgency) or time trends, to a large extent.

3.4   Mass Media

Crowdfunding campaigns have been increasingly covered by online mass media and news websites. Broadcasting to a large audience could increase public awareness about the campaign and therefore increase the number of potential funders. During our data collection period, we periodically collected information about any news coverage of each campaign by major online news websites. We developed a web-crawler application to collect every news article related to GofundMe campaigns within Google News. While at least ten articles referring to different GofundMe campaigns were published every day, few of them referred to campaigns that were included in our data (note that all the 590 campaigns in this study were created on a specific day and several hundreds of such campaigns are created every day). These campaigns were excluded from our collection. We also removed one campaign that was mentioned in a live news channel.

3.5   Offline Word of Mouth

Besides online word of mouth (social media) and news articles (mass media) about a campaign, offline word of mouth could increase awareness about the campaign, and therefore affect both contributions and online word of mouth. To control for this effect in our model, we assumed that some potential funders or advocates would have heard about a campaign through offline word of mouth and, then visit the campaign web page to donate and/or share a post about the campaign on social media. Translating this into our fixed effect model, offline conversations during a time period $t-1$ affect both contribution and sharing in the next period, t. This assumption is realistic if the length of each time period is small. Therefore, both the number of posts on social media about a campaign and contributions in time period t might be affected with (unobserved) offline word of mouth about the campaign at time $t-1$. To control for this effect, and consistent with the literature (Thies et al. 2014 and Hong et al. 2015), we modeled the effect of online word of mouth in previous periods on the contributions in the current period.

4 Empirical Model

In order to control for unobservable, time-invariant campaign characteristics (including inherent urgency of campaign's goal and fundraiser's experience), we collected multiple observations for each campaign (at different periods). This allows us to model the changes in campaigns' observable time-variant characteristics while controlling for unobservable time-invariant factors. Table 1 define and describe the main variables in the constructed panel data.

Table 1. Variables Definition

	Variable	Definition
1.	Contribution	Number of contributions to a campaign made in a particular period
2.	Facebook	Number of Facebook shares about a campaign in a particular period
3.	Twitter	Number of tweets posted about a campaign in a particular period
4.	Goal	A campaign's goal (in US Dollars)
5.	Percent	Percentage of a campaign's goal achieved
6.	Day	Number of days passed from beginning of a particular campaign
7.	Time	Time_1 through Time_4 representing four 6-hour periods in a day

We, then, employed a fixed effect (within effect) panel data model (Eq. 1) to estimate the model coefficients. The estimators in fixed effect modeling, which is equivalent to introducing a dummy variable for each campaign, are not affected by the heterogeneity of the time-invariant characteristics of the campaigns.

$$Contribution_{it} = Facebook_{it-1} + Twitter_{it-1} + News_{it-1} + Percent_{it} + Day_t + Time_t + Contribution_{t-1}$$

$$(1)$$

As noted above, both contributions (the dependent variable) and different WOM measures (the main independent variables) are skewed heavily to the right. Using the date in the original scale is likely to violate the assumption of normality of regression residuals and produce unreliable test statistics. A possible solution is to use log-transformed variables (Hong et al. 2015). However, the treatment for the log transformations of zero values (such as using $\log(x + c)$ for a constant c) are quite arbitrary. In cases where variables obtain the value of zero in a large number of observations, as in our data, the treatment of zero values in the transformation might affect the results, significantly.

In order to mitigate this problem, we modeled the presence (or non-presence) of events instead of the counts of them. More specifically, we can avoid the issue of dealing with skewed data containing many zeros by converting them to "presence data," binary variables indicating the presence or lack specific events (Fletcher, 2005). We then employ a fixed-effect logit model to estimate the parameters of the model. Therefore, Eq. 1 models the presence of any contributions to campaign i at time t as a function of the presence of WOM and mass media about the campaign in the previous period, as well as the progress of the campaign in terms of days passed (Day$t$), and the percentage of the goal funded (Percent$t$). In addition, we controlled for any possible reinforcement or substitution effects by including contributions in the previous period (Contribution$it-1$).

5   Preliminary Results

We use three different fixed effect models to test and estimate the effect of various factors on contribution. First, we estimated the model using the original (non-normalized) data, following Thies et al. (2014). Second, we transformed both contribution and WOM factors, following Hong et al. (2015). Third, we used "presence data" transformation, as explained above. Consistent with Thies et al. (2014) and Hong et al. (2015), all three models suggest a positive effect of Facebook shares on contribution. In contrast to these studies, both our original data and presence data models suggest a significant effect of tweets on contribution. We believe the disagreement between our results and the previous studies is mainly due to the differences in data collection or transformation. As discussed previously, failing to observe the effect of tweets in previous studies might be due to the issue of many zeros in the log-transformed model. Interestingly, our log-transformed model shows no effect of tweets on contribution as the log transformed model in Hong et al. (2015). Due to the real-time nature and short-lived effect of tweets, we believe that the data collection intervals should be small to capture these effects. Since all the main factors in our model were constructed based on periods of 6 h, we were able to observe the immediate effect of tweets on contributions. Previous studies collected and analyzed data on a daily basis.

Based on our preliminary logit model, on average, the presence of any Facebook shares in current period increases the odds of receiving at least one contribution in the next three periods by 21, 31, and 31%, respectively. Tweets, on the other hand, increases the odds of at least one contribution only in the next period, by almost 44%. In addition, the results suggest a significant negative effect of both time and percentage funded on contributions.

6   Conclusion

We empirically demonstrated the ways in which WOM affects contributions to personal crowdfunding campaigns, after controlling for other factors, including the campaigns' progress and previous contributions. Sharing posts on either Facebook or Twitter resulted in a significant increase in the odds of receiving funds during a short period thereafter. Both time and percentage funded negatively affected contribution. The effect of previous contributions, different forms of WOM (e.g. tweets vs retweets) and the characteristics of the person sharing the post (e.g. the number of his followers) are not reported in this manuscript due to space limitations. In addition, we have not reported our robustness checks, namely the assessment of multicollinearity, robustness to outliers, and alternative estimators including dynamic panel data models.

References

Agrawal, A.K., Catalini, C., Goldfarb, A.: Some Simple Economics of Crowdfunding. National Bureau of Economic Research, No. w19133 (2013)

Bénabou, R., Tirole, J.: Incentives and Prosocial Behavior. National Bureau of Economic Research, No. w11535 (2005)

Castillo, M., Petrie, R., Wardell, C.: Fundraising through online social networks: a field experiment on peer-to-peer solicitation. J. Public Econ. **114**, 29–35 (2014)

Fletcher, D., MacKenzie, D., Villouta, E.: Modelling skewed data with many zeros: a simple approach combining ordinary and logistic regression. Environ. Ecol. Stat. **12**, 45–54 (2005)

Guy, B.S., Patton, W.E.: The marketing of altruistic causes: understanding why people help. J. Consum. Mark. **6**, 19–30 (1989)

Harbaugh, W.T.: What do donations buy? A model of philanthropy based on prestige and warm glow. J. Public Econ. **67**, 269–284 (1998)

Hong, Y., Hu, Y., Burtch, G.: How does social media affect contribution to public versus private goods in crowdfunding campaigns? In: Thirty-Sixth International Conference on Information Systems, Fort Worth, Texas (2015)

Lehner, O.M.: Crowdfunding social ventures: a model and research agenda. Venture Capital **15**, 289–311 (2013)

Lu, C.-T., Xie, S., Kong, X., Yu, P.S.: Inferring the impacts of social media on crowdfunding. In: Proceedings of the Seventh ACM International Conference on Web Search and Data Mining, pp. 573–582 (2014)

Moqri, M., Bandyopadhyay, S.: Please share! Online word of mouth and charitable crowdfunding. In: Proceedings of Twenty-Second Americas Conference on Information Systems (2016)

Sargeant, A.: Charitable giving: towards a model of donor behaviour. J. Mark. Manage. **15**, 215–238 (1999)

Saxton, G.D., Wang, L.: The social network effect: the determinants of giving through social media. Nonprofit Voluntary Sect. Quart. **43**(5), 850–868 (2014)

Snipes, R.L., Oswald, S.L.: Charitable giving to not-for-profit organizations: factors affecting donations to non-profit organizations. Innovative Mark. **6**, 73–80 (2010)

Thies, F., Wessel, M., Benlian, A.: Understanding the dynamic interplay of social buzz and contribution behavior within and between online platforms–evidence from crowdfunding. In: Thirty-Fifth International Conference on Information Systems, Auckland (2014)

Predicting Web User's Behavior:
An Absorbing Markov Chain Approach

Sungjune Park$^{(\boxtimes)}$ ⓘ and Vinay Vasudev ⓘ

The Belk College of Business, The University of North Carolina at Charlotte,
9201 University City Blvd., Charlotte, NC 28223, USA
{supark,vkvasude}@uncc.edu

Abstract. We develop a novel predictive modeling framework for Web user behavior with web usage mining (WUM). The proposed predictive model utilizes sequence-based clustering, to group Web users into clusters with similar Web browsing behavior, and absorbing Markov chains (AMC) in order to model Web users' navigation behavior. Clustering facilitates the prediction of Web users' navigation behavior by identifying groups of Web users showing similar browsing patterns. The use of AMC allows calculation of transition probabilities and absorbing probabilities at any given time of active user sessions, and thus leads to a better Web personalization and a more effective online advertising outcome. This research will also provide a performance evaluation framework along with the proposed model and suggest a WUM system that can improve ad placement and target marketing in a website.

Keywords: Web mining · Predictive analytics · Markov chain · Clustering

1 Introduction

Within the past two decades, many organizations have begun implementing value-added services on the Web to gain competitive advantages by attracting loyal customers. In order to make the Web more user-friendly for individuals and create long-term relationships with them, companies now realize that providing personalized web services is crucial. In addition, online advertising has become major sources of revenue for many business organizations with large websites with heavy traffic. Web usage mining (WUM) allows for extraction of knowledge about such personalization and smart online advertising and thus leads to better Web experiences [1]. WUM refers to analyzing the data generated by the Web users' interactions with the Web including Web server access logs, user queries, and mouse-clicks, in order to extract patterns and trends in Web users' behaviors. A growing interest in the business use of 'intelligent' Web, also known as, Web 3.0, and social networking sites accentuate the importance of utilizing such patterns and trends for the purpose of creating effective marketing tools as well as enhancing user experiences on the Web.

This research proposes a novel predictive modeling framework for web user behavior with web usage mining. The proposed predictive model utilizes sequence-based clustering, to group Web users into clusters with similar Web browsing behavior, and absorbing Markov chains (AMC) in order to model Web users' navigation behavior.

© Springer International Publishing AG 2017
M. Fan et al. (Eds.): WeB 2016, LNBIP 296, pp. 170–176, 2017.
https://doi.org/10.1007/978-3-319-69644-7_17

Clustering facilitates the prediction of Web users' navigation behavior by identifying groups of Web users showing similar browsing patterns. Sequence-based clustering enables full consideration of sequential activities on the Web such as page visits or content views, which is a significant improvement over the usual practice of considering the frequency of visits to web pages. The Markov model has also been shown to be effective in predicting Web user's sequential navigation patterns. The use of AMC allows calculation of absorbing probabilities as well as transition probabilities at any given time of active user sessions, and thus leads to a better Web personalization and a more effective online advertising outcome.

Therefore, the main objectives of this research are (1) to develop a Web user behavior prediction model that integrates sequence-based clustering and the use of the AMC, (2) to provide a performance evaluation framework along with the proposed model, and (3) to suggest a WUM system that can be used to improve online advertising and target marketing, which are important subsets of Web personalization applications and revenue management for business.

2 Literature Review

Many data analytics techniques such as clustering, classification, association rules, sequence pattern analysis, and dependency modeling have been applied to Web server logs [2, 3]. Past research on the use of cluster analysis to identify Web user groups has primarily focused on clustering web users based on the frequency of their page visits. Cluster analysis based on sequences of Web navigation remains a relatively undeveloped area [4–7]. This is probably due to the dimensional complexity resulting from sequential data representation. A recent study [8] showed that sequence-based clustering effectively finds meaningful groups that share common interests and behaviors of Web users. Another interesting finding is that many studies discuss the need for a dynamic and adaptive clustering system, where clustering adapts to the continuous flow of new inputs in real-time. But, only a few studies [9, 10] presented implementations of dynamic clustering systems. The emergence of Web 3.0 and continued enrichment in Web 2.0 are expected to empower existing Web personalization applications. The significance of this research is thus widely acknowledged because it provides improvements those applications through the knowledge discovered from sequence-based and dynamic clustering methods.

Also, while the efficacy of using Markov chains to model Web user navigation behavior has repeatedly been stated in many past WUM studies [2, 11–13], the WUM literature rarely addresses the integration of a Markov chain-based prediction model and cluster analysis. Developing such an integrated model, along with an evaluation framework that enables a systematic comparison of the various techniques, can help to close obvious gaps in the WUM literature. Web user cluster formation research should address the development of clear clustering methodology that can handle sequential information effectively and efficiently. The methodology should be tested with real data in such a way that practical implications are highlighted, for example, the effectiveness of integration of user clusters in a Web user behavior prediction model.

Unlike the existing Markov models focusing primarily on link prediction, i.e., predicting the next page visit by a Web user, the proposed research goes beyond link prediction and provides a model that can predict additional important behaviors, such as remaining time of a user session and probability of purchase. The introduction of AMC in the proposed model allows a calculation of the expected number of visits to transient states before absorption (remaining session time) and the probability of entering a specific absorbing state (purchasing a specific product, downloading a specific file, etc.) from a pool of absorbing states. This additional information may help business organizations with a large website with heavy traffic optimally time their online advertising. It can potentially help online retailers better predict their product sales by analyzing the web user clusters and absorbing probabilities. To the best of our knowledge, there has not been any research discussing the applicability of AMC in such revenue management domains.

3 Research Methodology

The methodological contribution of this research mainly lies on the predictive modeling framework that integrates sequence-based clustering with a stochastic model of web user navigation behavior, i.e., absorbing Markov chain.

An absorbing Markov chain [14] is a special type of Markov chain, whose transition matrix has the following canonical form.

$$P = \begin{pmatrix} Q & R \\ 0 & I \end{pmatrix}$$

where I is an r-by-r identity matrix, 0 is an r-by-t zero matrix, R is a nonzero t-by-r matrix, and Q is an t-by-t matrix. The first t states are transient states and the last r states are absorbing states. Hence, Q represents transition probabilities between web pages. The chain will never leave the absorbing state it enters. These absorbing states may represent Web browser closing or log-out from the user session, where no subsequent user behaviors are observed. Analyzing these absorbing states is important because they oftentimes represent a completion of web user's task, such as purchasing a product, searching for specific information, consuming contents for entertainment, socializing, etc.

For an absorbing Markov chain P, the fundamental matrix N [14] is given as

$$N = I + Q + Q^2 + \ldots = (I - Q)^{-1}$$

The entry n_{ij} of N gives the expected number of times that the process is in the transient state s_j if it is started in the transient state s_j. Therefore, the expected number of transitions before the chain is absorbed is calculated as the ith element of $t = Nc$ where c is a column vector all of whose entries are 1.

Finally, the absorption probabilities can be calculated from the following matrix **B**.

$$\mathbf{B} = \mathbf{NR}$$

The element b_{ij} of **B** is the probability that the chain will be absorbed in the absorbing state s_j if it starts in the transient state s_i.

In order to accomplish the research objectives mentioned above, we first adopt a general sequence-based clustering method developed in Park et al. [6]. Using the replicated clustering approach, a widely-accepted method for comparison of clustering algorithms, we then investigate whether the identifying the correct clusters, i.e., groups of Web users who follow the same Markov process, enhances predicting the length of remaining user sessions and the absorbing state. A series of experiments will be conducted in order to determine whether prediction performance is affected by factors such as sequence representation scheme or clustering method as well as by other factors such as the number of actual Web user clusters, the number of pages, similarity between clusters, minimum session length, the number of user sessions, and the number of clusters to form.

Since we are interested in absorbing behavior of the Web users, the Markov chain will define absorbing states, which can vary from simply a departure from the website to a final outcome of a browsing activity (e.g., purchase and leave, leave without a purchase, etc.), to become an AMC. Assuming Web users share some browsing patterns, each cluster formed from sequence-based clustering is represented as one Markov chain. In other words, each cluster representing a class of web users with similar navigation patterns has its own transition matrix. In order to construct the transition matrix for a cluster k, denoted by $\mathbf{P}(k)$, each transition probability $p_{ij}(k)$ of the cluster is estimated from the frequencies of transition from state i to state j in the training data set.

After getting transition matrices for each cluster, a prediction of the remaining session length can be obtained in two steps as shown in Fig. 1. First, an incomplete session is classified into one of the clusters formed by sequence-based clustering. The classification algorithm uses cluster centroids, which can be obtained from the training

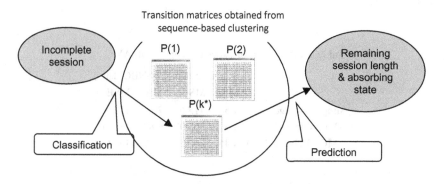

Fig. 1. Overview of the integrated navigation prediction model

set where *k*-means variants are applied as a sequence-based clustering algorithm. The best fit cluster for a given incomplete sequence is simply the cluster that has the minimum distance to the sequence. The prediction step involves matrix calculations as described above. The calculation of expected number of transitions before absorbing and absorbing probabilities from at a given transient state in AMC leads to the prediction of desired outcomes.

4   Results and Conclusions

In order to test the effectiveness of the AMC-based prediction model, we first tried predicting the remaining session length only, and the preliminary results are summarized in Table 1. We generated 10,000 synthetic Web user sessions from three dissimilar Markov chains, where navigation behaviors three Web user clusters are described with 15 states (page or page categories). The user sessions are then divided into a training set (80%) and a test set (20%) for performance evaluation. Finally, the overall performance of the prediction model was measured by the mean absolute error (MAE) and mean squared error (MSE) of the prediction for remaining session length (number of page visits) on the test set. The result of this experiment indicates that using AMC adds more predictive power compared to using the simple average of expected session length. Identifying correct clusters and using the corresponding AMC further improved the predictions of remaining session length.

We also applied the clustering and AMC-based prediction methodology to a publicly available large Web log at http://archive.ics.uci.edu/ml/ and presented the result in Table 1. The data set is generated from Internet Information Server (IIS) logs for msnbc.com for the entire day of September 28, 1999. User sessions are recorded at the level of page category. We used the first 10,000 user sessions and calculated MAE and MSE after finding three clusters. The result from the real data set also suggested that the AMC approach may improve the predictive power. However, whether clusters further improve the prediction was inconclusive. This may be in part due to the fact that we do not know the right number of Web user clusters. Another possible reason is that Web user behavior follows a higher-order Markov chain, i.e., Web users remember past visits and make navigation decisions based on the past visits and the page the user currently browse.

Table 1. Prediction of expected length with AMC

	Synthetic data		MSNBC data	
	MAE	MSE	MAE	MSE
Simple average	4.53	41.66	7.25	153.88
AMC (one MC)	4.5	36.29	6.87	136.86
AMC (3-cluster MC)	4.45	36.15	6.97	135.91

While it is not possible to discuss the full expected result, the proposed research will clearly be the first one to present the use of AMC for web user behavior and its outcome prediction. As the Web continues to grow exponentially, there are more and more opportunities to discover and utilize useful information and knowledge through web usage mining. This research-in-progress makes it possible the conversion of these opportunities into successes. It is expected to contribute to the web usage mining and web user behavior prediction literature by

- modeling Web user behavior based on AMC,
- providing effective sequence-based clustering method for identifying Web user clusters,
- offering novel predictive modeling and performance evaluation frameworks through systematic experiments and sensitivity analysis,
- suggesting practitioners new tools for Web personalization and revenue management, and
- suggesting a WUM system that can improve ad placement and target marketing.

Acknowledgment. This research is funded in part by a Belk College Summer Research grant from the Belk College of Business, UNC Charlotte.

References

1. Ho, S.Y., Bodoff, D., Tam, K.Y.: Timing of adaptive web personalization and its effects on online consumer behavior. Inf. Syst. Res. **22**(3), 660–679 (2011)
2. Facca, F.M., Lanzi, P.L.: Mining interesting knowledge from weblogs: a survey. Data Knowl. Eng. **53**(3), 225–241 (2005)
3. Pierrakos, D., Paliouras, G., Papatheodorou, C., Spyropoulos, C.D.: Web usage mining as a tool for personalization: a survey. User Model. User-Adap. Inter. **13**(4), 311–372 (2003)
4. Kim, Y.: Weighted order-dependent clustering and visualization of web navigation patterns. Decis. Support Syst. **43**(4), 1630–1645 (2007)
5. Kumar, P., Krishna, P.R., Bapi, R.S., De, S.K.: Rough clustering of sequential data. Data Knowl. Eng. **63**(2), 183–199 (2007)
6. Park, S., Suresh, N.C., Jeong, B.-K.: Sequence-based clustering for Web usage mining: a new experimental framework and ANN-enhanced K-means algorithm. Data Knowl. Eng. **65**(3), 512–543 (2008)
7. Shahabi, C., Banaei-Kashani, F.: Efficient and anonymous web-usage mining for web personalization. INFORMS J. Comput. **15**(2), 123–147 (2003)
8. Hung, Y.-S., Chen, K.-L.B., Yang, C.-T., Deng, G.-F.: Web usage mining for analysing elder self-care behavior patterns. Expert Syst. Appl. **40**(2), 775–783 (2013)
9. Borges, J., Levene, M.: Generating dynamic higher-order Markov models in web usage mining. In: Jorge, A.M., Torgo, L., Brazdil, P., Camacho, R., Gama, J. (eds.) PKDD 2005. LNCS, vol. 3721, pp. 34–45. Springer, Heidelberg (2005). doi:10.1007/11564126_9
10. Da Silva, A., Lechevallier, Y., de Carvalho, F., Trousse, B.: Mining web usage data for discovering navigation clusters. In: Proceedings of 11th IEEE Symposium on Computers and Communications, ISCC 2006, pp. 910–915. IEEE (2006)

11. Cadez, I., Heckerman, D., Meek, C., Smyth, P., White, S.: Model-based clustering and visualization of navigation patterns on a web site. Data Min. Knowl. Discov. **7**, 399–424 (2003)
12. Deshpande, M., Karypis, G.: Selective Markov models for predicting web-page accesses. ACM Trans. Internet Technol. **4**(2), 163–184 (2004)
13. Sarukkai, R.R.: Link prediction and path analysis using Markov chains. Comput. Networks **33**(1), 377–386 (2000)
14. Grinstead, C.M., Snell, J.L.: Introduction to Probability. American Mathematical Soc., Providence (2012)

Examining Customer Responses to Fake Online Reviews: The Role of Suspicion and Product Knowledge

Jie Ren[1(✉)], Pinar Ozturk[2], and Shoufu Luo[3]

[1] Fordham University, New York City, USA
jren11@fordham.edu
[2] Duquesne University, Pittsburgh, USA
ozturkp@duq.edu
[3] City University of New York, New York City, USA
sluo2@gradcenter.cuny.edu

Abstract. Online reviews constitute a central element in modern word-of-mouth communication and can strongly influence customer purchase intention. However, customers may be also aware of that these tools can be manipulated or counterfeited, and suspicion upon the review authenticity may affect its influence. The objective of this paper is to examine the effects of suspicion about fake reviews on the effectiveness of reviews in influencing customers' purchase intentions. The results of our empirical study show that customers who are suspicious of review authenticity find the reviews less convincing and reverse their likelihood to acquire the product. Furthermore, it holds true regardless of prior knowledge of the product.

Keywords: Fake reviews · Suspicion · Product knowledge · Information Processing theory · Experiment · Movies

1 Introduction

People increasingly rely on online reviews to make their purchase decisions [11, 13]. As a result, for businesses, these communication venues create opportunities for manipulation of public opinions for various purposes. There has been considerable anecdotal evidences that the presence of fake online reviews is endemic in many industries [4]. In addition to confirmed evidence, customers also hear rumors that companies may hire workers to post fake positive reviews for their own products or post fake negative reviews for their competitors' products. [25].

Thus, the general public is aware and, in many cases, suspects that online communication venues can be used for deception and review manipulation to propagandize goods and services. This can post a destructive impact on trust in online communities that collect, aggregate and present product reviews. Fearing the consequences of fake reviews, scholars have been studying this problem for years. For example, computer scientists have put considerable efforts into detecting online fake reviews from different perspectives [e.g. 16]. However, only a few have studied the topic of fake reviews from

© Springer International Publishing AG 2017
M. Fan et al. (Eds.): WeB 2016, LNBIP 296, pp. 177–184, 2017.
https://doi.org/10.1007/978-3-319-69644-7_18

the perspective of customer behaviors [2]. On the one hand, review-hosting platforms, such as Yelp, have incorporated fake review detection algorithms into their systems, but in most scenarios, those platforms can at best warn customers about the possible existence of fake reviews [19]. On the other hand, techniques of generating fake reviews have advanced to produce reviews nearly identical to real ones, posting new big challenges to existing fake review detection systems [21]. Since the amount of manipulation in online reviews is unknown and sources of manipulation are usually unconfirmed, most of the customers are left to their own discretion and intuition to assess the accuracy of the information and decide which reviews to believe. Thus, it is important to understand how customers behave in response to the existence of fake reviews. This paper is among the first attempts to explore customers' responses to products that may have received fake reviews. To study these responses, we conducted an online experiment involving 1,440 subjects from Mechanical Turk. Specifically, we examined how suspicion and prior product knowledge affect people's purchase intention given products that may have fake reviews.

2   Fake Review and Suspicion

People by nature are social and can be subject to others' opinions. Therefore, inevitably, when moving from offline to online, this human nature also exhibits itself [17]. As encountering with anonymity online, users rely on different cues to assess the creditability of the information source and further to establish a trust towards this source. Consequently, the concept of trust has received a lot of research attention and has been identified as a key driver for the success of electronic commerce [20]. Scholars studying e-business as well as electronic word-of-mouth (eWOM) and online communities particularly paid attention to the concept of trust and its determinants such as credibility of the source, first impressions (e.g. website design) [26] and repeated interactions [28].

Among this literature on online trust, the topic of fake reviews has been less studied, even though it is emerging and increasingly prevalent. We define fake reviews as reviews (often online) that are manipulated in order to persuade customers in the way the reviewers intended, regardless of the truth about the commented product. It is difficult to study this particular topic. First, detecting fake reviews is questionable. Computer scientists have been devoted to enhance the accuracy of the algorithms that can detect fake reviews [e.g., 16]. However, the state-of-the-art techniques cannot guarantee the accuracy. In other words, there are always false positives or false negatives. Second, despite the broad use of hidden paid posters (water armies) by companies, there is currently no systematic way to identify them - largely because these water armies mostly work "underground" and no public data is available to study their behavior [9].

Even so, online users seem to have developed their own decision-making mechanisms to identify fake reviews. They rely on their suspicion. In other words, based on certain cues (e.g., particular wordings, length, tones in text etc.), customers question the credibility of a particular review and discount the value of it. Despite these observations, past studies that examine trust online have not generally included conditions where customers are actively suspicious of the information they receive. The current study examines such conditions to determine whether suspicion affects customers' purchase intentions and whether prior product knowledge would buffer the generalized suspicion that customers feel when receiving information online.

3 Theoretical Model

Psychology literature has defined and studied suspicion, as a frequently-experienced mental state and often related to "an event whose occurrence is detrimental to the individual's welfare – i.e., in reference to a malevolent or undesired event" [12, p. 267]. As a result, suspicion makes people hesitate to act [14]. Prior literature studying suspicion has drawn two competing conclusions. On the one hand, researchers attributed customer suspicion to the situational factors (such as the information being displayed). For example, McKnight and Chervany listed seven different vendor interventions that contribute to trusting beliefs and intention, including privacy policy, linking to other sites, reputation building and customer interaction [20]. On the other hand, suspicion is also attributed to the carriers' own characteristics (e.g., personalities, education backgrounds etc.). Some people are found to be more inclined to trust than others [20].

To study the effects of suspicion in the online review community and its impact on customer behaviors, this paper builds upon Information Processing theory that include both situational factors (information) and individual characters (capacity to process information). Information Processing theory equates humans to computers that input information and output behaviors. In other words, people process information they receive from the environment and accordingly form short-term and long-term memories. And then cognitive processes (e.g., perception, thinking reasoning etc.) emerge and later affect their behaviors. People also constantly develop, as more information is received and analyzed; and their capacity, i.e., the information that they have stored before, can influence how they analyze additional information.

Similarly, when shopping online, online reviews can be considered as the input information for processing, given customers' own knowledge about a particular product, customers may question the credibility of these reviews to different extents and consequently make their own purchase decisions. Based on Information Processing theory, we developed our theoretical model as below (Fig. 1).

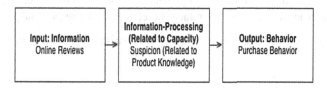

Fig. 1. Theoretical model built on Information Processing theory

4 Hypotheses Development

What happens when customers suspect that some of the reviews they are reading might be fake? Social psychology studies suggest that suspicion can decrease belief in false information. Fein et al. has reported that people's readiness to discount information increases if they are suspicious of the information and the motives underlying its dissemination [15]. Applying the literature on suspicion to fake reviews, we argue that suspicion of fake review presence will reduce belief in these review content. Moreover, as negative and positive reviews do not have the same impact on sales [3] and customer's evaluations [19] the effect of suspicion is likely to differ for positive versus negative reviews. If positive reviews are thought to be fake, their perceived utility would decrease and their positive effect on customers' purchase intentions would weaken, and hence customers would be less likely to purchase. In contrast, when negative reviews are thought to be fake, their detrimental effects on customers' purchase intentions would weaken and hence, customers would be more inclined to purchase. Therefore, we propose our first hypothesis.

H1. Suspicion that fake reviews exist will reverse people's tendency to acquire commented products.

Since product knowledge has often been regarded as one of the important moderators that determine purchase decisions [7], here we consider product knowledge as the moderator that affects the impact of suspicion. Product knowledge is based on memories [6] and customers who are knowledgeable about a product tend to have strongly established memory structure for that product [1]. Such memory structure does not only affect customer's information search behavior [6], but also affect his/her decision-making processing. Since they already have enough information to make an accurate purchase decision, they would devote less effort to obtaining additional product information or other customers' product evaluations [5]. Moreover, customers who are knowledgeable about the product are also found to be less susceptible to persuasion and external claims [8, 18].

In the context of evaluating fake reviews, prior research suggests that having prior knowledge on a topic helps people to recognize incoming information as either true or false [23]. Prior knowledge also allows for faster processing and checking for inconsistencies between incoming information and the knowledge retrieved from memory [27]. If relevant product knowledge is available, individuals use it for fast and efficient monitoring of incoming information and hence are able to reject false assertions efficiently when they have relevant background knowledge. Therefore, we posit that relevant product knowledge would reduce the effect of fake review suspicion.

H2. Having relevant product knowledge would reduce the effects of fake review suspicion.

5 Method

To test our hypotheses, we conducted an online experiment involving 1,400 subjects. Since manipulation of public opinions is often heard to be associated with the movie industry, we used movies to study online behaviors in response to fake reviews [25].

The experiment is based on a 2-by-3-by-2 full factorial design with unequal sample sizes. The first factor relates to online review valence (positive versus negative) and the second relates to the suspicion variable (no knowledge of fake reviews versus suspicion that fake reviews exist versus no suspicion that fake reviews exist). Thirdly, to measure prior product knowledge, we also asked subjects whether they have heard of the movie. In addition, we measured a few variables to control for their potential influence on purchase intention. Specially, prior literature suggests that online review texts are either high arousal or low arousal and the arousal in online review texts affects sales [22], therefore we controlled for the arousal of the reviews. We also measured frequency of movie watching to control for the impact of demand on purchase intention. In addition, we asked subjects for gender and age, and used six different movies to control for the impact of movie genre (six movies represented action, romance, thriller, animation, comedy and sci-fi genres respectively). We asked subjects "how likely do you want to watch the movie?" on a 7-point-likert scale to measure subjects' intended purchase behavior.

6   Procedure and Data Collection

We used Mechanical Turk as the platform to collect data from subjects [24]. Each of the subjects was asked to read the title and plot description of a movie as well as three related review texts. Then each was asked his/her likelihood to watch this movie. By not revealing the poster of the movie and not revealing the actors or actresses in the movie, we purposely controlled for the impact of celebrity on movie purchase decisions. In the conditions where fake reviews are possible, we constructed the stimuli by saying "there are rumors saying fake review(s) were posted about this movie in general. And it is unsure if these displayed reviews are fake." Given this stimulus, we defined suspicious subjects as those who thought at least one review text was fake, and defined non-suspicious ones as those who thought none of the review texts was fake. Specifically, after subjects have indicated their likelihood to watch the movie, we asked subjects to select which review text(s) is (are) fake (to identify suspicious subjects). Subjects can select "none" if they think none of the review texts are fake (to identify non-suspicious ones). In contrast, in the no fake review condition, we didn't post any message relating to fake reviews. In addition, across all conditions, to measure the majority of control variables, we asked subjects whether they have heard of the movie (to measure product knowledge), how frequently they have watched movies in general (to measure movie-watching demand), and what are their age and gender (to measure age and gender).

Review texts were extracted from movie sites including Rotten Tomato, iMDB and Netflix to measure the valence and emotionality for each displayed movie. For each movie, all three review texts were displayed randomly as either positive and emotional, or positive and rational, or negative and emotional, or negative and rational.

In summary, subjects were randomly assigned to one of the sub-conditions that were constituted by review valence (2 values), review emotionality (2 values), fake review variable (2 values: fake versus no fake reviews) and movie genre (6 values) to read reviews and indicate their purchase decisions. Each subject only participated in the study once.

In total, 1,400 subjects participated in our study and after removing incomplete responses, 1290 responses remained in the data pool. All the subjects were United-States based workers who had approval rates, as tabulated by Amazon, higher than 95%: 49% of those workers were female and 51% were male. Subjects were between 18 and 72 years of age (average 35 years old), consistent with demographic results from other studies [24].

7   Results

H1 was supported. We ran two t-tests with unequal sample sizes and compared the likelihood of watching the movie in the 2 (positive versus negative reviews) by 3 (no knowledge of fake reviews versus suspicion that fake reviews exist versus no suspicion that fake reviews exist) full factorial design (Fig. 2). We found that when people are suspicious that fake reviews exist, they are more likely to watch a negatively

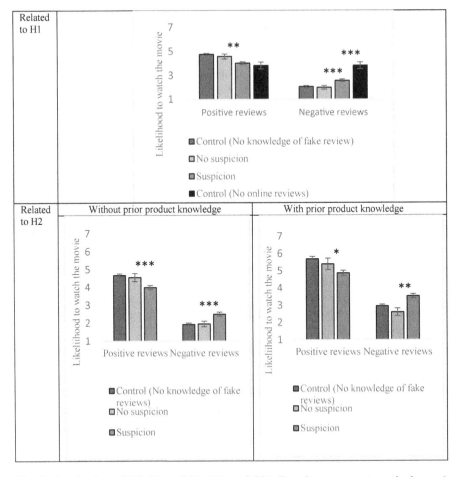

Fig. 2. Results (*p < 0.05; **p < 0.01; ***p < 0.001; Error bars represent standard errors)

commented movie and less likely to watch a positively commented movie. Interestingly, the suspicion that fake reviews exist has weakened the negative impact of negative reviews proportionally more than the positive impact of positive reviews. The significant interaction between online review valence and the suspicion variable (p < 0.001) in ANOVA results also supported this hypothesis.

H2 was not supported. We conducted t-tests and compared the movie watching likelihood of subjects who know about the movie versus those who do not. We found that regardless of prior product knowledge and online review valence, subjects with suspicion about the presence of fake reviews tended to alter their likelihood to watch the movie compared to those with no suspicion. Moreover, the ANOVA result suggests no significant interactions among review valence, the suspicion variable and product knowledge. This suggests that H2 was not supported (Fig. 2).

8   Discussion and Conclusion

Fake reviews have become epidemic and have largely affected customers' online behaviors. We believe our work proposes a new behavioral perspective of studying fake reviews, which contributes to the ongoing research of online reviews in general. Specifically, this paper partially validates information processing theory in the domain of processing fake reviews. However, human cognitive bias also exhibits. We found that regardless of prior knowledge of the displayed product, facing reviews that are possibly fake, customers have the same tendency to be suspicious and accordingly reverse purchase decisions. Moreover, studying suspicion opens up a possible good venue to study emerging phenomena in online review communities such as fake reviews. Our analysis reflects the first stage in our larger research project of this research topic. In subsequent papers, we will incorporate real-setting data and study other products than only movies [10].

References

1. Alba, J.W., Hutchinson, J.W.: Dimensions of consumer expertise. J. Consum. Res. **13**, 411–454 (1987)
2. Bambauer-Sachse, S., Mangold, S.: Do consumers still believe what is said in online product reviews? A persuasion knowledge approach. J. Retail. Consum. Serv. **20**, 373–381 (2013)
3. Basuroy, S., Chatterjee, S., Ravid, S.A.: How critical are critical reviews? The box office effects of film critics, star power, and budgets. J. Market. **67**, 103–117 (2003)
4. BBC: Samsung probed in Taiwan over 'fake web reviews' (2013). http://www.bbc.com/news/technology-22166606
5. Bloch, P.H., Sherrell, D.L., Ridgway, N.M.: Consumer search: an extended framework. J. Consum. Res. **13**, 119–126 (1986)
6. Brucks, M.: The effects of product class knowledge on information search behavior. J. Consum. Res. **12**(1), 1–16 (1985)
7. Celsi, R.L., Olson, J.C.: The role of involvement in attention and comprehension processes. J. Consum. Res. **15**, 210–224 (1988)
8. Chakravarty, A., Liu, Y., Mazumdar, T.: The differential effects of online word-of-mouth and critics' reviews on pre-release movie evaluation. J. Interact. Market. **24**, 185–197 (2010)

9. Chen, C., Wu, K., Srinivasan, V., Zhang, X.: Battling the Internet water army: detection of hidden paid posters. In: Advances in Social Networks Analysis and Mining (ASONAM), pp. 116–120. IEEE Press (2013)

10. Chiu, C.M., Wang, E.T., Fang, Y.H., Huang, H.Y.: Understanding customers' repeat purchase intentions in B2C e-commerce: the roles of utilitarian value, hedonic value and perceived risk. Inf. Syst. J. **24**, 85–114 (2014)

11. Dellarocas, C., Zhang, X.M., Awad, N.F.: Exploring the value of online product reviews in forecasting sales: the case of motion pictures. J. Interact. Market. **21**, 23–45 (2007)

12. Deutsch, M.: Trust and suspicion. J. Confl. Resolut. **2**, 265–279 (1958)

13. Duan, W., Gu, B., Whinston, A.B.: Do online reviews matter?—An empirical investigation of panel data. Decis. Support Syst. **45**, 1007–1016 (2008)

14. Fein, S.: Effects of suspicion on attributional thinking and the correspondence bias. J. Personal. Soc. Psychol. **70**, 1164–1184 (1996)

15. Fein, S., Morgan, S.J., Norton, M.I., Sommers, S.R.: Hype and suspicion: the effects of pretrial publicity, race, and suspicion on jurors' verdicts. J. Soc. Issues **53**, 487–502 (1997)

16. Feng, S., Xing, L., Gogar, A., Choi, Y.: Distributional footprints of deceptive product reviews. ICWSM **12**, 98–105 (2012)

17. Hennig-Thurau, T., Gwinner, K.P., Walsh, G., Gremler, D.D.: Electronic word-of-mouth via consumer-opinion platforms: what motivates consumers to articulate themselves on the internet? J. Interact. Market. **18**, 38–52 (2004)

18. Herr, P.M., Kardes, F.R., Kim, J.: Effects of word-of-mouth and product-attribute information on persuasion: an accessibility-diagnosticity perspective. J. Cust. Res. **17**(4), 454–462 (1991)

19. Li, H., Chen, Z., Liu, B., Wei, X., Shao, J.: Spotting fake reviews via collective positive-unlabeled learning. In: IEEE International Conference on Data Mining (ICDM), pp. 899–904 (2014)

20. McKnight, D.H., Chervany, N.L.: Conceptualizing trust: a typology and e-commerce customer relationships model. In: Proceedings of the 34th Annual Hawaii International Conference on System Sciences, p. 10 (2001)

21. Mukherjee, A., Kumar, A., Liu, B., Wang, J., Hsu, M., Castellanos, M., Ghosh, R.: Spotting opinion spammers using behavioral footprints. In: Proceedings of the 19th ACM SIGKDD International Conference on Knowledge Discovery and Data Mining, pp. 632–640. ACM Press (2013)

22. Ren, J., Nickerson, J.V.: Online review systems: how emotional language drives sales. In: Twentieth Americas Conference on Information Systems (2014)

23. Richter, T., Schroeder, S., Wöhrmann, B.: You don't have to believe everything you read: background knowledge permits fast and efficient validation of information. J. Personal. Soc. Psychol. **96**, 538 (2009)

24. Ross, J., Irani, L., Silberman, M., Zaldivar, A., Tomlinson, B.: Who are the crowdworkers?: shifting demographics in Mechanical Turk. Paper presented at the CHI 2010 Extended Abstracts on Human Factors in Computing (2010)

25. Sasaki, K.: (2016). http://www.bleachbypass.com/batman-v-superman-fake-imdb-ratings/

26. Schlosser, A.E., White, T.B., Lloyd, S.M.: Converting web site visitors into buyers: how web site investment increases consumer trusting beliefs and online purchase intentions. J. Market. **70**, 133–148 (2006)

27. Schroeder, S., Richter, T., Hoever, I.: Getting a picture that is both accurate and stable: situation models and epistemic validation. J. Mem. Lang. **59**, 237–255 (2008)

28. Sillence, E., Briggs, P., Harris, P., Fishwick, L.: A framework for understanding trust factors in web-based health advice. Int. J. Hum.-Comput. Stud. **64**, 697–713 (2006)

An Exploration of Public Reaction to the OPM Data Breach Notifications

Rohit Valecha[1], Eric Bachura[1], Rui Chen[2($\boxtimes$)], and H. Raghav Rao[1]

[1] University of Texas at San Antonio, San Antonio, TX, USA
rohit.valecha@utsa.edu,
ericbachura@gmail.com, mgmtrao@gmail.com
[2] Iowa State University, Ames, IA, USA
ruichen@iastate.edu

Abstract. With the number of data breaches swelling, people are likely to not respond adequately and ignore the breach notifications altogether. Ignorance of breach notifications creates a perfect storm of cyclical outrage and apathy that criminals can use to their advantage. In this research-in-progress paper, we explore public reactions to breach notifications for addressing two research questions: (1) with more and more information related to the breach, do people become apathetic towards that breach? (2) at what point do people simply tune out information related to the breach? The results of the sentiment analysis show that public express anxiety when there is a fear of being affected by the breach, and public express anger when there are lack of measures to safeguard the data. Sadness is the most strongly expressed emotion in response to the severity of the breach. After the public has received sufficient details about the event, they start to tune out information related to the breach event.

Keywords: OPM breach · Twitter microblog · Public reaction · Emotions · Sentiments

1 Introduction

According to the Identity Theft Resource Center, there occurred a total of 5,810 data breaches with a total of 847,807,830 individual records stolen between 2005 and 2015 (IRTC 2016). Data breaches have been reported by companies such as JP Morgan Chase, Target, Neiman Marcus, LinkedIn, EBay, Adobe, Home Depot, and Sony Pictures. Each of these data breaches represent an incident where private data has been reviewed, stolen, or used by unauthorized parties. This data may include personally identifiable information (PII), personal health information (PHI), or other information protected with the purpose of limiting access to authorized users. Data from the Privacy Rights Clearinghouse shows that external hacking is the leading source of breaches, followed by insider disclosure, physical loss, and lost or stolen devices (Williamson 2015). According to CSID, a leading provider of global identity protection and fraud detection technology, the financial, educational, and government sectors have the highest number of breaches reported (CSID 2015). Compromised data can be exploited by criminals to commit crimes affecting a victim's identity and finances.

M. Fan et al. (Eds.): WeB 2016, LNBIP 296, pp. 185–191, 2017.
https://doi.org/10.1007/978-3-319-69644-7_19

Vulnerable consumers who do not take protective measures may incur losses that may have otherwise been preventable. Anecdotes have shown that consumers may not respond adequately to data breaches. Examples include a lack of post-breach vigilance, no alteration in their behavior, and neglecting to subscribe to fraud protection services (O'Farrell 2014; Perlberg 2014). Some may ignore alarms, alerts, or breach notifications. A 2014 Ponemon Institute report found 32% of consumers surveyed "ignored the notifications and did nothing" when alerted to a data breach involving their information (PonemonInstitute 2014). The same report also found that 55% of consumers surveyed had taken no steps to protect themselves from future identity thefts. A YouGov BrandIndex report found that consumers turned numb to breach incidents (Marzilli 2013). Criminals can take advantage of people if they do not care about the breach notifications, a result of which they will be continued targets of criminal activity. Indeed, it is clear that investigating public's reactions to breaches is of paramount importance.

Most of the work studying public reaction to breaches has focused on a macro-level understanding of people's response to multitude of data breaches as the number of breaches have multiplied from 2011 to 2016 (the year of mega breaches). The idea behind the macro-level investigation of public reaction is that the more people are confronted with breaches, the less likely they are to care about other breaches. To date, however, there is little micro-level understanding of public reaction in the aftermath of an individual data breach. With more and more information related to the breach, do people become apathetic towards that breach? At what point do people simply tune out information related to that breach? In this research-in-progress paper, we explore negative emotions (anger, anxiety, fear and sadness) in the aftermath of the breach notifications within the context of a single breach event. The contributions of this research-in-progress paper are two-fold: First, it determines the pattern of emotional responses to breach notifications. Second, it investigates if people will start treating breach notifications apathetically.

The study is organized as follows. In the next section, we review the background of public reaction to breaches and specify the context of the research. We then explore the data about a recent breach and conduct an analysis of emotions related to the breach. Subsequently we analyze the breach related data to identify two critical dimensions that need to be considered in studying people's reaction to breaches. Finally, we end this study with the conclusion.

2 Background

OPM Breach and Twitter Microblog. For the purposes of this paper, we focus on the April 2015 data breach of the U.S. Office of Personnel Management (OPM). In this incident, personnel data – full name, birth date, home address, and social security number – of millions of Federal employees was stolen. It was a large data breach in the recent past and its sheer number of victims allows us rich data in understanding publics' reaction to breaches. This preliminary study lays the foundation for subsequent projects (e.g., model extensions or victims and non-victims comparisons).

The widespread use and the rapid growth of microblogging services have resulted in new mechanisms for people to share their emotions regarding specific events. Microblogging services have afforded broadcasting information to the public in real time thereby enabling people to express their emotions without any time or location restrictions (Lee et al. 2016). In this way, microblogging services can be considered as useful tools for corporate entities and government organizations alike in understanding people's reaction in the aftermath of data breaches. In this paper, we focus on Twitter messages in the aftermath of OPM breach.

Public Reaction to Breaches. Industry reports have suggested that consumers demonstrate a lack of interest in breach incidents with increasing occurrences of breach events (Humphries 2014). Public reaction to breaches is a result of their prior experience, both direct and indirect, with data breaches. It remains relatively consistent across situations and may act as an important cognitive bias that distorts people's sense making in the aftermath of a breach event. Lee et al. (2016) argue that emotional factors can inform public's reaction to events. A decision maker's apathy may be explained by his or her insensitivity to data breach notifications, which results from an over exposure to frequent media reports on similar incidents in the past (Lazzarotti 2014). When one becomes drained by hearing about data breach events, he or she tends to lose interest in breach events.

To best of our knowledge, most of the studies investigating breach events have ignored the micro-level view that examines publics' reaction to individual aspects of the breach events. In order to examine the breach event from a micro-level view, we investigate people's emotional state in response to the data breach notifications. Following Lee et al. (2016), who suggest that three negative emotions – anger, anxiety, and sadness – help in understanding the people's response to a crisis event, we explore negative emotions (anger, anxiety and sadness) in the aftermath of the breach notifications within the context of OPM breach event. Anger is denotes an uncomfortable response to a grievance (Videbeck 2013); Anxiety is the fear caused by being aware of danger (Lee et al. 2016); Sadness is the opposite of happiness (Lee et al. 2016).

3   Methodology

OPM Timeline. While the adversarial access to OPM's network dates back to mid-2012, the breach was only detected in mid-2014 when OPM was notified by a third party. The adversary managed to obtain an elevated access to the OPM's network, and successfully exfiltrated personnel information, including fingerprints. OPM released public notifications in June 2015. The data we utilize in this research-in-progress paper provides the key events in the timeline post the public notification of the OPM breach (see Table 1).

Data Collection. We purchased data following the OPM data breach using third party vendor. Twitter provides three APIs to enable researchers and developers to collect data, namely STREAMING, REST and SEARCH APIs. Satisfying user specified

Table 1. Timeline of OPM breach

Date	Event	Importance
6/4/15 – First time #opmhack concept enters public sphere – Begin data collection		
6/4/15	First public notification	First time #opmhack concept enters public sphere
6/8/15	H. comm. brief confirming lost data	News reports relay briefing: public now aware of data significance
6/12/15	Fed union: hackers took SSNs of everyone	News report claiming SSNs of all federal employees taken
6/16/15	OPM Director Archuleta acknowledgement	OPM Director finally acknowledges breach
7/9/15	Press release confirming breach magnitude (21.5 million affected)	Magnitude of impact revealed to be larger than thought
7/10/15	OPM director archuleta resigns	Director resigns (1 day after major press release)
7/31/15 – End data collection		

filtering criteria (based on keywords, location, language, etc.), streaming API is used to get tweets and their corresponding user's data in real time, REST API is used to get the data in select historical time period, and search API provides data on relevant searches on Twitter. For this research, the purchased tweets were collected from Twitter microblogs through the REST APIs using the keyword #opmhack. We requested purchased data from the time when OPM publicly announced the breach (June 4th 2015) till two months after the announcement (July 31st 2015). This time window allowed us to gather 18,764 tweets that provided information during the time with the greatest propensity for public reaction to data breach. Table 2 provides the descriptive of the data collected. Figure 1 shows a distribution of the tweets collected.

Table 2. Descriptive data

Month	Tweets (N)	Percent (%)
June 2015	9018	48.06%
July 2015	9746	51.94%

Sentiment Analysis. In order to investigate people's reaction to OPM breach notifications, we conducted sentiment analysis of the tweet messages. Sentiment analysis allows us to identify the expression of sentiments in the text messages (Nasukawa and Yi 2003). Sentiment analysis combines natural language process and text analysis to extract emotion information from the text (Stieglitz and Krüger 2011). We used a text analysis software, Linguistic Inquiry and Word Count (LIWC), for analysis of anger, anxiety and sadness expressed within the tweets. LIWC uses a psychometrically

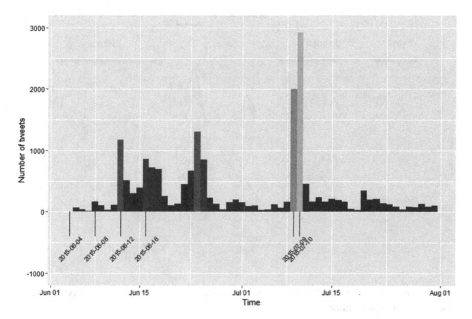

Fig. 1. Distribution of the tweets collected

validated dictionary to measure the inherent emotions in the text (Lee et al. 2016). LIWC is a popular tool for sentiment analysis and has been used in research to extract emotions from plain tweets (Pennebaker et al. 2015; Tausczik and Pennebaker 2010)

4   Emotional Reaction to OPM Data Breach

In this section, we present analysis results to answer the two research questions: First, we investigate the research question – "With more and more information related to the breach, do people become apathetic towards that breach?" From Fig. 2, we can see that when the information about suspicious traffic on OPM network was confirmed by US-CERT and when the federal union claimed that SSNs of all federal employees were stolen, people expressed anxiety because of the fear of being affected by the event (event 1). As OPM Director acknowledged the compromise of background data, public expressed their anger that stems from the lack of measures in place to safeguard the personnel data (event 2). When OPM released a public statement that 21.5 million personnel were affected, people experienced sadness for the extent of damage the breach caused (event 3).

Second, we investigate the research question – "At what point do people simply tune out information related to that breach?" After the press release, people obtained sufficient detail about the incident, which is visible in the less volatile patterns (somewhat stable patterns) of anger, anxiety and sadness (event 4). In this phase, people reduced their reactions to the OPM breach by tuning out information related.

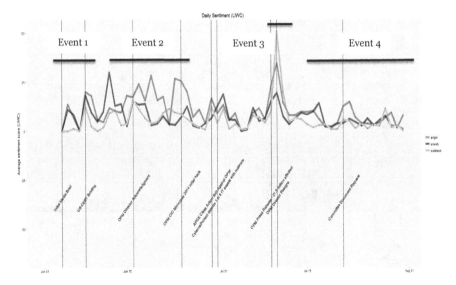

Fig. 2. Public reaction to OPM breach over the timeline (red: anger; green: anxiety; and blue: sadness) (Color figure online)

5 Discussion and Conclusion

The existing research, for the most part, has been silent on online public's reactions in response to data breaches (Chakraborty et al. 2016). Our paper makes a twofold contribution. First, it determines the emotional responses to breach notifications. Second, it investigates if people will start treating breach notifications apathetically. In the case of breach events, people expressed anxiety in the early stages when there is a danger from the event. In addition, public also showed anger for OPM's incapability to safeguard personnel information. In reaction to the sizeable loss of data (21.5 million records), sadness was the most strongly expressed emotion. In this way, the results confirm that expressed emotions reflect the characteristics of the breach event.

Findings of our study offer directions to the practitioners and policy makers in conceiving plans that promote protection in the aftermath of data breach incidents despite challenges. This preliminary study lays the foundation for future studies targeting at uncovering more complex issues. We plan to test the model among non-victim populations and to include other constructs that may better explain this research phenomenon.

This study has the following limitations. First, the information about OPM hack was available much before it was publicly announced. Second, in the timeline, we have only focused on those events that were available to public. Third, we only focus on one microblogging platform, Twitter. This could affect the generalizability of the results. One possibility of future work would be to look at positive emotions and contrast its patterns with the negative emotions. Yet another future option would be to breakdown the timeline to accommodate events that were not necessarily available to public in the

OPM context. Finally, the emotional reactions over the timeline can be constructed for each user and then compared it with the social reaction (as reported in this paper).

Acknowledgement. This research has been funded by NSF under grants 1554373, 1554480, and 1651060. The usual disclaimer applies.

References

Chakraborty, R., Lee, J., Bagchi-Sen, S., Upadhyaya, S., Rao, H.R.: Online shopping intention in the context of data breach in online retail stores. Decis. Support Syst. **83**, 47–56 (2016)

CSID: Data Breaches Pose a Threat across All Industries. CSID, Austin (2015)

Humphries, D.: Public awareness of security breaches: industry view 2014. In: Software Advice. Gartner (2014)

IRTC: Data Breaches, Identity Theft Resource Center. San Diego (2016)

Lazzarotti, J.: Report says Russian hackers stole 1.2 billion usernames and passwords, but don't let "breach fatigue" take hold. In: Workplace Privacy, Data Management and Security Report. Jackson Lewis P.C, White Plains (2014)

Lee, J., Rehman, B.A., Agrawal, M., Rao, H.R.: Sentiment analysis of Twitter users over time: the case of the Boston bombing tragedy. In: Sugumaran, V., Yoon, V., Shaw, Michael J. (eds.) WEB 2015. LNBIP, vol. 258, pp. 1–14. Springer, Cham (2016). doi:10.1007/978-3-319-45408-5_1

Marzilli, T.: Target perception falls after data breach. In: YouGovBrandIndex. YouGov plc, London (2013)

Nasukawa, T., Yi, J.: Sentiment analysis: capturing favorability using natural language processing. In: Proceedings of the 2nd International Conference on Knowledge Capture, pp. 70–77. ACM (2003)

O'Farrell, N.: Data Breach Fatigue: Consumers Pay the Highest Price. CreditSesame.com, Mountain View (2014)

Pennebaker, J.W., Boyd, R.L., Jordan, K., Blackburn, K.: The Development and Psychometric Properties of LIWC2015. UT Faculty/Researcher Works (2015)

Perlberg, S.: Do consumers have data breach fatigue? In: WSJ.com. New York (2014)

PonemonInstitute: The Aftermath of a Mega Data Breach: Consumer Sentiment. Ponemon Institute (2014)

Stieglitz, S., Krüger, N.: Analysis of sentiments in corporate Twitter communication–A case study on an issue of Toyota. Analysis **1**, 1–2011 (2011)

Tausczik, Y.R., Pennebaker, J.W.: The psychological meaning of words: LIWC and computerized text analysis methods. J. Lang. Soc. Psychol. **29**(1), 24–54 (2010)

Videbeck, S.: Psychiatric-Mental Health Nursing. Lippincott Williams and Wilkins, Philadelphia (2013)

Williamson, W.: Data breaches by teh numbers. In: Security Week. Wired Business Media (2015)

Behavior Theory Enabled Gender Classification Method

(Research in Progress)

Jing Wang[1(✉)], Xiangbin Yan[2(✉)], and Bin Zhu[3(✉)]

[1] Harbin Institute of Technology, Harbin, China
jingwangxr@gmail.com
[2] University of Science and Technology Beijing, Beijing, China
xbyan@ustb.edu.cn
[3] Oregon State University, Corvallis, USA
Bin.Zhu@bus.oregonstate.edu

Abstract. While it is crucial for organizations to automatically identify the gender of participants in product discussion forums, they may have difficulties adopting existing gender classification methods because the associations between the linguistic features used in gender classification models and gender type usually varies with context. This paper proposes and validates a framework for the development of gender classification that uses a more "data-driven" approach. The framework constantly extracts content-specific features from the discussions and could automatically adjust the features selected to accommodate the contextual changes in order to achieve better classification accuracy. It does not require any manual effort for model adjustment, which makes it easier for organizations to adopt.

Keywords: Text mining · Gender classification · Chinese gender lexicon · Mutual information · Support vector machine

1 Introduction

The advent of web 2.0 leads to the formation of many online communities that focus the discussions on a specific product type. For example, people post feedback about movies at imdb.com, discuss their experience with cosmetic products at makeupalley. com, or express their product opinions at epinion.com. And almost all the online retailing websites provide platforms for product discussions. These product specific communities not only provide a place for organizations to better understand customers' perception about their products, but also offer a channel for marketers to reach customer groups with related interest. Knowing the gender of participants will not only enhance the effectiveness of the effort of recommending related products/services to forum participants but also could help organizations better understand the gender difference existed in customer perceptions about their products. It thus is crucial for organizations to avail themselves with a gender classification tool that automatically identifies the gender of participants in product discussion forums.

© Springer International Publishing AG 2017
M. Fan et al. (Eds.): WeB 2016, LNBIP 296, pp. 192–200, 2017.
https://doi.org/10.1007/978-3-319-69644-7_20

However, although many gender classification methods have been proposed [1–4], the adoption of these methods, especially in the context of online product discussions, still poses a special challenge. Most existing gender classification studies focus on the improvement of classification accuracy. They usually used a list of pre-selected linguistic features with known polarized value for gender type as the independent variables to develop a gender classification model. The classification accuracy largely depends on the configuration of the selected linguistic feature list. However, the association between a linguistic feature and gender type is usually contextual, mediated by factors such as genres [5], authors' social networks [1], social classes [1], etc. A linguistic feature could be a gender marker under one circumstance but loses its discriminatory power in another [5]. More importantly, the gender difference in certain linguistic features could be reversed as the context changes [6]. Therefore, in order to conduct gender classification in online discussion communities, an organization needs a classification method to be able to automatically adjust the feature selection to accommodate the contextual changes.

The goal of this study is to propose a framework to guide the development of gender classification systems specifically for online product discussions. It therefore is not the intention of this paper to propose a classification method that leads to a higher classification accuracy compared with other existing methods. Instead, our goal is to develop a more operationally feasible approach that could be easily adopted. The rest of the paper is organized as follows. Section 2 presents the background of the study along with the development of hypothesis, followed by Sect. 3 that depicts the proposed architecture for the development of a gender classification system. Section 4 describes the experiment conducted for the hypothesis testing. And Sect. 5 provides discussions and summaries.

2 Background and Hypothesis Development

2.1 Motivation

The importance of knowing the gender of a given online participants is evidenced by many marketing studies investigating how the gender difference moderates the impact of marketing strategies. For example, men and women have been found to have very different perception about the risk associated with online transactions [7]; they put different weight on the same decision factors when making purchasing decisions [8], and they could react very differently to the same advertisement [9]. It therefore becomes apparent that a marketing strategy could be more effective if it also factors in the gender difference. And this would be possible only when marketers can automatically identify the gender of a given audience. Therefore, being able to automatically identify the gender of online community members becomes an important antecedent for organizations to take advantage of the existing insight on gender difference.

2.2   Gender Classification

Gender classification denotes to the automatic identification of the gender of the author for a given document. Coming from the computer science tradition, this stream of studies belongs to a larger research category of computational research that seeks to automatically identify the values of social variables of people such as age, gender, or region origins, etc. The research of gender classification was based upon the assumption that men and women write differently [10] in terms both of the communication topic and of the writing style. For example, for communication content, women tend to make mention of emotion and mostly in a self-derogatory way, whereas men are inclined to state opinions [11]. From the perspective of writing style, male language are regarded as direct, succinct, personal and instrumental, while female language is more indirect, more elaborate, and more affective [12].

More importantly, gender classification studies believe that above mentioned gender difference could be extracted and be aggregated to identify the gender of authors. The development of a gender classification system thus usually consists of three steps. The first step selects pre-defined features with known polarity values for different gender types as independent variables. The second step creates a classification model using the feature selected. And finally, the step 3 applies the model created over the input data to predict the gender of authors. Most existing studies are usually similar in classification algorithms, using support vector machines (SVM) [13], naïve Bayesian [2], or neural networks based algorithms, but are different from each other in the features they selected. In fact, most studies claim the configuration of different types of features they identified to be their major contributions [1–4].

2.3   Gender-Based Features

Content-free features and content-specific features are two types of commonly used features [4]. The content-free features measure people's habitual way of writing, also called writing style, which usually is independent of the writing content. Most existing gender classification studies provide indisputable evidence for the existence of strong associations between above features and gender type [4]. However, a specific association between a single content-free feature and the gender type has been shown to be contextual [1]. As a result, the selection of content-free features needs to be constantly adjusted to accommodate the contextual changes. And such adjustment usually requires manual effort.

On the other hand, content-specific features have been found to have better performance in author classification systems than content-free features [14]. Content-specific features capture the topic preferences of different gender type for a given subject domain. The topic preference is usually represented by a list of key words/phrases extracted from the training data set. The difference between a content-free feature and a content-specific feature lies in how the independent variables are selected. The former starts with a fixed set of variables believed to have strong associations with gender type, while the later starts with extracting variables with significant different values for each gender type. The contextual information thus has been automatically accommodated during the process of extracting content-specific variables from the data. Therefore even though the association between a content-specific feature and gender type is contextual, the adjustment of features selected could be automatically accomplished without any manual effort.

2.4 Hypothesis Development

When it comes to online discussion forums, it usually is not very difficult to collect hundreds of thousands data points for the development of gender classification model. But the conventional view that larger data set is always associated with better performance may not apply in this context. Because the discriminatory power of both types of features is contextual, it could be easily diluted by large data set. Not only a gender marker may change its discriminatory power when context changes, it may also reverse its association with gender classes, losing its discriminatory power completely [15]. A very large data set may cover several topics discussed by different groups of participants. As a result, the features that appear to be optimal to the entire data set may not stay optimal for a subset of the same data set. At the same time, many studies have found that the popularity of different discussion topics largely varies time [4]. Thus in the context of online discussion forums, it is not an unreasonable assumption that people who are participate in online discussions are more likely talking about the same topic in same time than people who participate in different time. Therefore, if we segment a discussion archive by time, each subset should have lower content diversity compared with the original large-scale archive. Consequently,

> *H1. Using a linguistic feature list and classification model locally built over the subset to conduct gender classification over the subset itself should lead to higher classification accuracy than using a linguistic list and classification model globally created over the original large-scale archive to conduct the gender classification over the subset.*

While larger training data size does not always lead to better classification performance, the size of the training data set should not be too small either. It usually requires the training data set to reach certain threshold in order for the discriminating power of gender markers to be apparent [16]. Consequently, the size of a training data set should be large enough for the gender markers to have discriminating power but small enough to keep the same context for the selected gender markers to be effective. Therefore,

> *H2. When dividing very large data set into subsets, there is an optimal data set size for each data segment that leads to the highest classification accuracy.*

> *H3. After the size of a data segment reaches its optimal size, the classification accuracy will be negatively associated with the size of that data segment.*

3 Proposed System Architecture

We proposed the system architecture for the development of a gender classification system for online product discussion forums. It consists of four sub-systems, word segmentation, feature optimization, model creation, and gender classification. The main idea behind this proposed architecture is to restart the entire process when the size of newly generated data set reaches a threshold. Extract and select features only from the new data set. A new model using the features selected as independent variables will be trained using only the data from the new data set.

The *word segmentation sub-system* represents the content of an input document with a list of key words, phrases, or concepts. We adopted the ICTCLAS (Institute of Computing Technology, Chinese Lexical Analysis System) technique for the development of the word segmentation sub-system. It is reported that ICTCLAS' segmentation speed is around 500 KB/s and its precision has reached 98.45% [17]. As one of the best Chinese lexical analyzers [18], ICTCLA is widely used in the field of Chinese word segmentation [3]. We then removed words with extremely low frequencies after running ICTCLAS because it is very difficult to examine the link between an uncommonly used word and a gender type.

The *feature optimization sub-system* extracts content-specific features from the data set. The outcome of this sub-system is a Chinese gender lexicon that contains words and associated weights. Negative value indicates a "male" word, while a positive one indicates "female" word. A word with value around zero means that this is a neutral word; and there is no difference in word usage between men and women. Among the words with negative values, the smaller the weight value is, the more popular the word is among male writers. And for words with positive values, the larger the weight value is, the more likely the word will be picked up by women.

After word segmentation, we calculated the mutual information (MI) between a word and each of the two gender classes. MI is a concept in information theory that measures the statistical relationship between two variables. The larger is the absolute value of MI, the more significant is the relationship between the two variables. We adopted the formula for MI calculation from [19] and listed the formula as follows:

$$I(F, C_i) = \log \frac{p(F|C_i)}{p(F)p(C_i)} \tag{1}$$

Here F is a word and C_i is one of the gender classes. $p(F \wedge C_i)$ is the probability of F appearing in C_i in the collection, while $p(F)$ and $p(C_i)$ are the probabilities of word F and class C_i appearing in the collection, respectively. For a word, two types of MI were calculated, with one for its relationship with the female class and the other for its relationship with male class. Then we calculated the weight value of each word through female MI minus male MI. After calculating words' weights, we selected words whose weight values are larger than a threshold that is measured as 25 percent of the maximum absolute value of the difference between female MI and male MI.

The *model creation sub-system* constructs a classification model over the training data whenever the lexicon is updated. The system indexes a document with the words in the lexicon, integrating the conventional TF/IDF approach with the weight values in the lexicon. The modified TF-IDF formula expresses as following [20]. In formula (2), $w(t, d)$ is defined as the weight of item t in document d, $tf(t, d)$ expresses the number of times that the word t appears in the document d, N is the total number of documents in the collection, N_t denotes to the number of documents which contain the term t, and w_t is the weight of word t in lexicon. The selected words using the information gain (IG) method along with its weighted values are then be used as the independent variables for the creation of classification model. The system described utilized SVM to create a classification model.

$$w(t,d) = \frac{tf(t,d) \times \log(\frac{N}{N_t} + 0.01) \times w_t}{\sqrt{\sum_{t \in d} [tf(t,d) \times \log\left(\frac{N}{N_t} + 0.01\right) \times w_t]^2}} \tag{2}$$

The *gender classification sub-system* applies the model created to predict the gender of the authors. Again, this sub-system represents each document using the same method as that used for model creation.

4 Hypothesis Testing

4.1 Testbed

The testbed data includes product discussions on cell phone downloaded from tianya. cn, the most popular online discussion forum in China with at least 80 million user accounts. We chose this product domain because it is one of the most popular product domains discussed at this website. The time period of our data covered from July 20th in 2004 to November 9th in 2008. The demographic information of a reviewer could be found from his/her profile. A review was not used for feature optimization and model creation if its reviewer did not indicate his/her gender. Finally, we obtained 37,643 cell phone reviews, including 10,987 female reviews and 26,656 male reviews.

4.2 Hypothesis Testing

In order to test the above hypothesis, we segmented data chronologically according to different sizes of training data sets, investigating the association between classification accuracy and the size of a training data set. In order to obtain a certain number of training data, we chronologically chose top half the certain number of both female and male reviews. The corresponding test data set consists of reviews that were posted no later than the later time between female and male training data. Removing the reviews appeared in test data set, the next round of gender classification starts with the same process, and so on. The size of training data set shifts from 100 to 1,000 reviews with step length of 100 reviews, 1,500 to 10,000 reviews with step length of 500 reviews, and 15,000 to total reviews with step length of 5,000 reviews. The size of training data will not extend if half size is larger than the number of total female or male reviews. Figure 1 shows the accuracy changes of gender classification with the size of the training data set.

   Figure 1 indicates that the gender classification with training data set that is a subset of the whole data set has higher accuracy than that with training data set that is the whole data set. For cell phone reviews, the accuracy over 1,000 training reviews is higher than that over total reviews as training data (0.778 > 0.750). A paired-samples t test was performed and the result indicates that the difference between the accuracy of

Fig. 1. The accuracy changes of gender classification with the size of the training data set

gender classification over a subset as training data (1,000 phone reviews) and that over total reviews as training data is significant ($P < 0.05$). In other words, using a linguistic feature list and classification model locally built over the subset to conduct gender classification over the subset itself should lead to higher classification accuracy than using a linguistic list and classification model globally created over the original large-scale archive to conduct the gender classification over the subset. H1 is supported.

From Fig. 1, we infer that training data with 1,000 reviews for cell phone has the peak accuracy. We conducted paired-samples t tests for comparing the peak accuracy with the accuracy over other different size of training data set. The results indicate that training data with 1,000 reviews has the highest classification accuracy and the accuracy is significantly higher than that of other size of training data set ($P < 0.05$). Hence, H2 is also supported. There is an optimal data set size for each data segment that leads to the highest classification accuracy when dividing very large data set into subsets.

A correlation test was performed to figure out the relationship between training data size and the corresponding gender classification accuracy after training data segment reaches the optimal size. As shown in Table 1, the correlation between classification accuracy and training data size is -0.428 ($P < 0.05$). It indicates that the classification accuracy is significantly negatively related with the size of training data set. This means that hypotheses H3 is accepted at a significant level of 0.05, meaning that the classification accuracy will be negatively associated with the size of that data segment after the size of a data segment reaches its optimal size.

Table 1. Correlation test between training data size and gender classification accuracy

		Accuracy of gender classification
Training data size	Pearson Correlation	-0.428
	Sig. (2-tailed)	0.047
	N	22

5 Summary and Discussion

This paper proposes and validates a framework for the development of gender classification system in the context of online product discussion forums. Unlike most existing studies that start the classification with a list of preselected features, the framework proposed utilized a more "data-driven" approach that constantly extracts content-specific features from the discussion content. It could automatically adjust itself to accommodate the contextual changes in order to achieve better classification accuracy. The framework does not require any manual effort for model adjustment, which makes it easier for organizations to adopt. A serial of evaluation studies have been conducted to validate the performance of the framework.

Acknowledgments. This work is partly supported by the National Natural Science Foundation of PRC (Nos. 71531013, 71490720, and 71401047).

References

1. Bamman, D., Eisenstein, J., Schnoebelen, T.: Gender in Twitter: styles, stances, and social networks. CoRR abs/1210.4567 (2012)
2. Mukherjee, A., Liu, B.: Improving gender classification of blog authors. In: Proceedings of the 2010 Conference on Empirical Methods in Natural Language Processing, pp. 207–217. Association for Computational Linguistics, Stroudsburg, PA, USA (2010)
3. Wei, X., Dong, P., Cui, G.: Automatic extraction of course ontology from chinese textbook. In: 2010 International Conference on Computational Intelligence and Software Engineering. IEEE (2010). doi:10.1109/CISE.2010.5677020
4. Yang, J., Leskovec, J.: Patterns of temporal variation in online media. In: Proceedings of the Fourth ACM International Conference on Web Search and Data Mining, pp. 177–186. ACM, New York, NY, USA (2011). doi:10.1145/1935826.1935863
5. Herring, S.C., Paolillo, J.C.: Gender and genre variation in weblogs. J. Sociolinguistics **10**(4), 439–459 (2006). doi:10.1111/j.1467-9841.2006.00287.x
6. Labov, W.: Principles of linguistic change, cognitive and cultural factors, vol. 3. John Wiley & Sons, Hoboken (2011)
7. Garbarino, E., Strahilevitz, M.: Gender differences in the perceived risk of buying online and the effects of receiving a site recommendation. J. Bus. Res. **57**(7), 768–775 (2004). doi:10.1016/S0148-2963(02)00363-6
8. Yang, C., Wu, C.C.: Gender differences in online shoppers' decision-making styles. In: Ascenso, J., Vasiu, L., Belo, C., Saramago, M. (eds.) e-Business and Telecommunication Networks, pp. 108–115. Springer, Dordrecht (2006). doi:10.1007/1-4020-4761-4_6
9. Doong, H., Wang, H.: Do males and females differ in how they perceive and elaborate on agent-based recommendations in Internet-based selling? Electron. Commer. Res. Appl. **10**(5), 595–604 (2011). doi:10.1016/j.elerap.2010.12.005
10. Savicki, V., Kelley, M.: Computer mediated communication: gender and group composition. CyberPsychol. Behav. **3**(5), 817–826 (2004). doi:10.1089/10949310050191791
11. Thomson, R., Murachver, T., Green, J.: Where is the gender in gendered language? Psychol. Sci. **12**(2), 171–175 (2001). doi:10.1111/1467-9280.00329
12. Mulac, A., Bradac, J.J., Gibbons, P.: Empirical support for the gender-as-culture hypothesis. Hum. Commun. Res. **27**(1), 121–152 (2001). doi:10.1111/j.1468-2958.2001.tb00778.x

13. Rao, D., Yarowsky, D., Shreevats, A., Gupta, M.: Classifying latent user attributes in Twitter. In: Proceedings of the 2nd international workshop on Search and mining user-generated contents, pp. 37–44. ACM, New York, NY, USA (2010). doi:10.1145/1871985.1871993

14. Martindale, C., McKenzie, D.: On the utility of content analysis in author attribution: the Federalist. Comput. Hum. **29**(4), 259–270 (1995). doi:10.1007/BF01830395

15. Eckert, P., McConnell-Ginet, S.: Constructing meaning, constructing selves: snapshots of language, gender and class from Belten high. In: Hall, K., Bucholtz, M. (eds.) Gender Articulated: Language and the Socially Constructed Self, pp. 469–507. Routledge, London and New York (1995)

16. Burger, J.D., Henderson, J., Kim, G., Zarrella, G.: Discriminating gender on Twitter. In: Proceedings of the Conference on Empirical Methods in Natural Language Processing, pp. 1301–1309. Association for Computational Linguistics, Stroudsburg, PA, USA (2011)

17. ICTCLAS: ICTCLAS features. http://www.ictclas.org

18. Bo, A., Peng, S., Xinming, T., Alimu, N.: Spatio-temporal visualization system of news events based on GIS. In: 2011 IEEE 3rd International Conference on Communication Software and Networks (ICCSN), pp. 448–451. IEEE (2011). doi:10.1109/ICCSN.2011.6014089

19. Yang, Y., Pedersen, J.O.: A comparative study on feature selection in text categorization. In: Proceedings of the Fourteenth International Conference on Machine Learning, pp. 412–420. Morgan Kaufmann Publishers Inc., San Francisco, CA, USA (1997)

20. Liu, W., Zhu, Y., Li, C., Xiang, H., Wen, Z.: Research on building Chinese basic semantic lexicon. J. Comput. Appl. **29**(10), 2875–2877 (2009)

Crowdfunding Platforms: The Role of Information Providers

Zhenhua Wu and Zhijie Lin$^{(\boxtimes)}$

School of Business, Nanjing University, Nanjing, China
w.nju@foxmail.com, mailtozjlin@gmail.com

Abstract. Despite the popular emergence of crowdfunding platforms, relevant research investigating the role of these platforms in crowdfunding markets still lags. In this paper, we present a model to study market incentives of crowdfunding platforms' optimal information reporting strategy when there exists uncertainty on projects' returns. We assume that platforms on the market are rational players, and they seek to stay on the market as long as possible as accurate information providers. We characterize platforms' equilibrium reporting strategies under different market conditions. Surprisingly, we find that under certain conditions, the potential competition from a new entrant gives the incumbent crowdfunding platform an incentive to bias the information on borrowers' risky projects. However, the uncertainty resolution provided by a third party (e.g., regulator, media) could reduce the incentive. Our findings contribute to the literature on crowdfunding by analyzing platform decisions, and offer policy implications for the regulations of crowdfunding markets.

Keywords: Crowdfunding platforms · Information provider · Financial intermediaries · Asymmetric information · Bayesian update · Perfect Bayesian Equilibrium

1 Introduction

"Yes, competition rewards the sharp and hardworking. But it also often compels them to keep the frontiers of subtle deception in view...The problem is that the promise of genuine 'unique information' comes with the reality of vulnerability to deception."

Robert J. Shiller [17]

Crowdfunding markets have experienced a rapid growth in many countries in the last few years. The global market valuation has increased from $880 million in 2010 to $16 billion in 2014, and is estimated to reach $90 billion by 2020 [6]. Witnessing the increasingly important role of crowdfunding in the economy, President Obama has decided to sign the "Jumpstart Our Business Startups (JOBS) Act" to legalize crowdfunding in the U.S. market in 2012. Although crowdfunding has been legalized and attracted many investors, the mechanism of investor protection is still thin. Crowdfunding platforms have been at the center of media attentions and policy debates.

© Springer International Publishing AG 2017
M. Fan et al. (Eds.): WeB 2016, LNBIP 296, pp. 201–214, 2017.
https://doi.org/10.1007/978-3-319-69644-7_21

Unlike traditional financial intermediaries that provide products such as bonds, crowdfunding platforms provide project information as their products or services. Given that information may directly influence investors' decisions, which implies that information may play a critical role in investor protection, we have to recognize the importance of information quality. The quality issue mainly comes from three aspects. First, crowdfunding platforms have an access to a wider variety and a larger set of information which makes them better informed than small investors, and the information accessed might completely depend on platforms' intrinsic characteristics (e.g., the experience of platform operators and staffs). Second, information provided by platforms could be biased (i.e., information diverges from the truth) as a result of platforms' strategic decisions. These two aspects may force small investors to be poorly informed of the true quality of these risky projects on platforms. Third, the quality of information is also affected by systematic risks in the macroeconomy, such as the policy uncertainty from the government. Additionally, the uncertainty from stock markets also generates a high risk to projects on platforms, because many small businesses on crowdfunding platforms may invest the funded money in stock markets due to the economies of scale. Consequently, these three aspects may lead to the classical incentive issue caused by asymmetric information, which is the central issue this research attempts to investigate.

In essence, our research question is: *How would market mechanisms, such as competition and government intervention, discipline crowdfunding platforms in the market?* In particular, we build a model to study the incentives of crowdfunding platforms choosing different information reporting strategies[1], under the assumptions that crowdfunding platforms are rational players and seek to stay on the market as long as possible as accurate information providers. As the quality of the risky projects is always hard to be observed or verified by small investors, investors' beliefs about quality would largely depend on the information provided by crowdfunding platforms. Apparently, investors would prefer a platform with better information (i.e., more information or more accurate information) to evaluate and choose the "best" projects for investments. However, whether a platform can obtain better information largely depends on its characteristics (e.g., the experience of platform operators and staffs). Therefore, platforms that cannot obtain better information may have incentives to misreport or conceal the observed information if the information indicates that the projects have low quality.

We find two notable results. The first result indicates that crowdfunding platforms have incentives to conform small investors' prior beliefs on the average quality of the risky projects by distorting observed information. To explain, we consider an observed signal which is more likely to generate reports that contradict the true average quality. If investors strongly believe that a state (e.g., high quality) is the true state of the quality of projects on a platform (e.g., the average quality of the risky projects or the actual quality of the prototype of the new product), they would expect more inaccurate

---

[1] Reporting strategies in this study can be broadly explained, for instance, slogans used in the advertisement, contents provided in the financial report, and even movie stars invited during new product release.

information than accurate information to contradict prior beliefs. For instance, a platform generates a report indicating that a crowdfunding project would produce a new smart phone which is better than the latest iPhone. If investors initially believe that this would be highly unlikely to happen, they will infer that the information in the report is distorted. Therefore, a platform that seeks to attract more investors would have incentives to generate a report to conform the investors by distorting or misreporting the evidence, even if the platform is aware that the evidence it observed is true. The more the investors favor the priors, the less likely the platform would generate reports contradicting those priors.

Our second result shows that when the market or a third party (e.g., regulator, media) offers small investors a chance to access to additional information to verify the true state of the quality of projects on the platform, then the platform's incentive to distort information will be weakened. This is because, if a platform on the market misreports the observed signal to conform investors' priors, it faces the risk that the truth would be revealed which then contradicts the report. This would be detrimental to the platform because investors might withdraw the investment and choose a new platform. As the chance of getting ex-post uncertainty resolution from a third party increases, the likelihood of misreporting in equilibrium decreases.

2 Literature Review

Our research is generally related to three streams of literature. First, our study is related to the growing literature on online peer-to-peer lending [e.g., 2, 3, 5, 13, 14]. Many researchers studied the role of information in online lending. For instance, Burtch et al. [4] examined the impact of borrowers' information control on their behaviors. Iyer et al. [10] explored whether lenders can infer soft information on Prosper.com. Moreover, Michels [14] investigated whether unverifiable information disclosure is associated with funding probability. In addition to information, social factors have also attracted some attention. For instance, Lin et al. [11] and Liu et al. [13] examined how friendships affect online lending. Lastly, researchers also attempted to study issues such as trust [5], home bias [12], and also the difference of funding decisions between crowds and experts [15]. Noteworthy, Hildebrand et al. [8] is the one most related to our research, which argued that the growth and viability of crowdfunding markets critically depend on the underlying incentives. However, most of these previous studies have either focused on factors related to the supply side or the demand side of a crowdfunding market, but overlooked the market of crowdfunding platforms itself. Thus, our research differs from prior literature by investigating how market incentives affect platforms' equilibrium actions in crowdfunding. Our model can be the building blocks for further research in crowdfunding markets.

Second, this research is related to the classical literature on how incentives shape agent's actions in financial markets when there exists information asymmetry. These studies, e.g., Gorton and Pennacchi [7] and Holmstrom and Tirole [9], modeled the interactions between the informed lenders and non-informed investors, and studied how to overcome the problem induced by adverse selection and moral hazard.

Third, our work is also closely related to the economics literature on herding on the priors. This literature studies how an agent's actions depend on prior beliefs of factors that may determine the agent's utility. Prendergast [16] found that this dependence could drive workers to bias the reports to match the information that managers have received, and valuable information could be lost in equilibrium. Moreover, Brandenburger and Polak [1] studied how shareholders' priors affect managers' decisions which may affect stock prices given that the managers are short run players. In contrast to this literature, the competition we model here in turn generates an inefficient equilibrium, and biased information could be transmitted by crowdfunding platforms, even if the platforms are long run players.

3   Model

We consider a model with two time periods denoted by $t \in \{1, 2\}$. In each period, there is a market with a continuum of homogeneous investors and a continuum of heterogeneous firms[2] (or borrowers or small businesses) which seek to raise funds from investors. Firms can only post their risky projects on a platform for investors to invest, but cannot get connected with investors directly outside platforms. In this paper, we focus on the market of platforms, so we do not model the interactions between firms and platforms. We simply assume that, given period $t \in \{1, 2\}$, a subset of firms are selected by a platform on the market. Firms are assumed to be heterogeneous on the quality of their risky projects. The average quality of risky projects on a platform is described by a binary state variable, $s_t \in \{H_t, L_t\}$, which is random and distributes according to $\text{Prob} \in \{s_t = H_t\} = \pi > 1/2$, where $H_t > L_t > 0$. Here, the states of the world are independent across time. Investors and platforms are assumed to have the common belief, π, on these state variables. In other words, the average quality of projects in each period only has two values. When the average quality of risky projects on the platform is high, the value would be H_t. When the average quality of risky projects on the platform is low, the value would be L_t.

Here, *quality* can be explained broadly. For instance, we can consider quality as the default rate of loans posted on a crowdfunding platform. The loans are assumed to only have two possible states (i.e., default or no default). Then, H_t indicates no default (high quality) and L_t indicates default (low quality).

To simplify the analysis, we only have two crowdfunding platforms on the market and assume that platform $j \in \{1, 2\}$ can be either an active type or passive type. We also assume that with probability $r_j \in (0, 1)$, where $j \in \{1, 2\}$, platform j will be an active type. If it is an active type, the platform would always perfectly observe a signal $\omega_t = S_t$, where $t \in \{1, 2\}$. If it is a passive type, an informative but imperfect signal will be observed:

$$\text{Prob}\{\omega_t = H_t | H_t\} = \text{Prob}\{\omega_t = L_t | L_t\} = q \in (\pi, 1)$$

---

[2] In this paper, we do not put any restriction on firms except that they are heterogeneous on the quality of the projects listed on the platform.

Meanwhile, platforms are aware of their own types with certainty. Here, we do not put any restriction on the relationship between r_1 and r_2. If $r_1 > r_2$, platform one faces a weak competitor which is ex-ante less likely than platform one to be an active type. Likewise, if $r_1 > r_2$, platform one faces a strong competitor.

We use $\theta \in \{\theta_1, \theta_2\}$ to denote the type of the platform on the market, where θ_1 indicates an active type and θ_2 indicates a passive type. Then we denote the passive platform's strategy conditional on its signal by $\sigma_{\omega_t}(x_t|\theta_2) = \text{Prob}(x_t|\omega_t, \theta_2)$. To reduce the issues caused by multiple equilibria, we also assume that an active platform always reports observed signal honestly, i.e., $\sigma_{\omega_t}(x_t = \omega_t|\theta_1) = 1$. Thus, only the passive platform can manipulate the observed signal by freely reporting either H_t or L_t. For simplicity, we denote the passive platform's strategy conditional on its signal by $\sigma(x_t|\omega_t) = \text{Prob}(x_t|\omega_t, \theta_2)$, and restrict our attention to the case under which $\sigma(H_t|H_t) \geq \sigma(H_t|L_t)$.

The interpretation of the passive type could be explained broadly. It could include platforms which are new to the crowdfunding industry and have limited competence. It could also capture the situation of a crowdfunding platform which has private interest to pursue. For instance, some peer-to-peer financial platforms use fake projects to collect money from investors and cash out all the money before the delivery date. However, the exact interpretation is not important to the results and discussions in this paper.

Since the investors are homogeneous, we assume that there is no coordination problem during the decision, i.e., all the investors would choose the same action in equilibrium. Therefore, we further assume that there is a representative investor to make every decision for these investors. This investor has one unit of asset to invest in each period. We can consider this as a proportion of income which will be used for investment by the investor. For this representative investor, he chooses whether or not to invest one unit asset on the platform in the first period. At the end of the first period, he also needs to decide whether to purchase more from platform one after observing the report x_t, or just switch to the new entrant and purchase one more unit there in the second period.

In period one, i.e., $t = 1$, platform one enters the crowdfunding market with a set of risky financial projects posted by firms which seek to raise funds. The average quality of projects posted on the platform is described by a random variable $s_1 \in \{H_1, L_1\}$, and the distribution is commonly observed. The investor decides whether to invest on the projects listed on the platform. If purchase takes place, the platform will investigate or monitor[3] the average quality of the projects and observe a signal related to the true state of the world. After that, the platform strategically reports $x_1 \in \{H_1, L_1\}$ to the investor. After observing the report, the investor updates beliefs on the type of the platform and the quality of the risky projects on this platform.

At the beginning of the second period, platform two enters the market with another set of risky financial projects. The average quality of the risky projects posted on this platform is described by a random variable $s_2 \in \{H_2, L_2\}$, and the distribution is also

---

[3] For instance, Kickstarter has an integrity team that employs complex algorithms and automated tools to identify and investigate suspicious projects.

commonly observed. The investor decides whether to keep the investment on platform one or switch to platform two with another one unit of asset realized in the second period. In each period, one unit of asset will be realized to the investor as an endowment for investment in this period. The platform, one or two, observes another signal related to the true state of the world in the second period and strategically reports $x_2 \in \{H_2, L_2\}$ to the investor. The outcome of investment is realized at the end of the second period.

Finally, we assume that, at the end of the first period, with probability $\rho \in [0, 1]$, the uncertainty of the average quality can be resolved before the investor making the purchase decision of the second period. We assume that, as long as the platform chooses the correct action, i.e., truthfully reporting the state of the world ($x_t = s_t$), the social welfare will be maximized accordingly.

For further analysis, we now introduce some notations. First, we define the states after which the investor would make purchase decision on the second period as $\mathbf{A} = \{H_1, L_1, \varnothing\}$. Then we use A_i to denote the i th element in set $\mathbf{A}$ which indicates the state of uncertainty resolution. Second, we denote $\mu(\theta_1 | x_t, A_i)$ as the investor's posterior on the platform on the market being an active type after observing the report x_t in period t.

In order to keep the analysis as simple as possible and without loss of generality, we assume that:

Assumption: If $\mu(\theta_1 | x_t, A_i) < r_2$, the investor would purchase from platform two. Otherwise, he would keep the investment on platform one.

Overall, the timeline of the game is as follows:

(1) nature determines the type of two platforms and the average quality of the risky projects, s_1, on platform one in the first period;
(2) platform one observes the first period signal s_1 according to the type and chooses report x_1;
(3) the investor updates beliefs and makes an investment decision based on the posterior beliefs on the true state of the average quality in period one;
(4) with probability $\rho \in [0, 1]$ the investor would observe the true state s_1 and update beliefs on the type of the platform and choose whether to keep the investment on platform one in the second period;
(5) if the investor withdraw the investment, platform two enters the market with another set of risky assets;
(6) nature determines the average quality of the risky projects, s_2, on platform two in the second period;
(7) the platform, one or two, on the market observes the second period signal s_2 according to the type and chooses report x_2;
(8) the investor updates beliefs and makes an investment decision based on the posterior on the true state of the average quality in period two;
(9) the return of investment is realized, and the payoffs are realized accordingly.

3.1 Preference

In this paper, we focus on the interactions between the platform's actions and the investor's purchase decision. Thus, we assume that the firms which seek to raise funds will not choose any actions. These firms' features are completely characterized by the state of the world, $s_t \in \{H_t, L_t\}$. Therefore, platform one's utility would be $U_1 = \mathbf{1}_{s_1} U + \mathbf{1}_{purchase} \beta \mathbf{1}_{s_2} U$, and platform two's utility would be $U_2 = (1 - \mathbf{1}_{purchase}) \mathbf{1}_{s_2} U$, where U is the platform's payoff in each period if it truthfully reports the observed signal,

$$\mathbf{1}_{purchase} = \begin{cases} 1 & \text{Purchse from platform} \quad 1 \\ 0 & \text{Purchse from platform} \quad 2 \end{cases} \text{and } \mathbf{1}_{s_t} = \begin{cases} 1 & \text{If} \quad x_t = s_t \\ 0 & \text{If} \quad x_t \neq s_t \end{cases}$$

The platform's outside options are normalized to zero. The motivation of this setup is as follows. We argue that the investor's belief on the state, e.g., the quality of the risky assets proposed by firms and the intrinsic characteristic of the platform, would influence the platform's reporting decision. If we allow the platform to have different payoffs for different states, the generated equilibrium prediction would depend on both investor's beliefs and the exogenously determined quality from investment. Then it would be hard to argue that the platform's misreporting is largely determined by the investor's beliefs. Thus, in order to strengthen our argument, we assume that truthful reporting always gives the platform the same positive expected payoff from investment. Then, we show that misreporting would still take place in equilibrium.

For the representative investor, we assume that his utility only depends on the reports from the platform and the average quality of the risky projects, i.e., $V = \mathbf{1}_{s_1} B + \beta \mathbf{1}_{s_2} B$. This specification implies that as long as the true state is the same as the reported signal, investors would get a positive quality B. Here, we assume the investor and the platform have the same discount factor. We also normalize the outside options of the investor to zero. Instead of specifying the pricing and market structure in details, we use the above setup because it is easy to check that the expected utility from investment is always positive. Therefore, a direct implication from this setup is that the representative investor would always invest one unit asset on one of the platforms in equilibrium in each period[4]. This would help us focus on the investor's decision on the platform selection at the end of the first period. It would further help us clarify the effects of market incentives.

4 Benchmark

4.1 Model Without Entrance Threat and Uncertainty Resolution

In this section, we repeat twice the second period of the model as our first benchmark. In this benchmark, we would analyze the market equilibrium of a monopoly platform

---

[4] If we specify a more general utility function for the investor, we can always construct an equilibrium under which the investment takes place on the investor side. However, this generalization would only complicate the analysis without generating any new insights.

without entrance threat and uncertainty resolution. That is, we have platform one on the market for two periods with a set of risky financial projects. The investor believes, with probability r_1, it is an active platform. The platform strategically reports the observed signal to the investor who will always purchase in equilibrium. The preferences are exactly the same as the ones specified in the previous section.

The key difference between this benchmark and the general model presented in the previous section is as follows. There is no competition from an entrant. Thus, a passive monopoly platform would have no incentive to pretend to be an active type. It can always get positive payoffs from truthfully reporting the observed signal. Then, we have the following result in equilibrium:

Proposition 1: In the monopoly market, for any type of platform, it would truthfully report the signal in any period.

Due to page limit, we would only explain intuition of the results. Omitted proofs are available upon request. This result implies that a passive monopoly platform would truthfully report the observed signal to the investor even if there is no way to get uncertainty resolved ex-post. Therefore,

(1) the incentive effect from information disclosure by a third party is limited in the monopoly case;
(2) the platform has no incentive to bias the report in order to pander the investor's interest.

4.2 Model in Two Special Cases ($\rho = 0$ and $\rho = 1$)

In this and the next part, we study the two special cases of the model presented in Sect. 4. The first one is the case with the assumption of $\rho = 0$, i.e., uncertainty about the true state of the world will not be resolved at the end of the first period. The second one is the case with the assumption of $\rho = 1$, i.e., uncertainty about the true state of the world is completely resolved at the end of the first period.

4.2.1 Uncertainty Is not Resolved ($\rho = 0$)
We first study the properties of the investor's posterior on the platform being an active type, $\mu(\theta_1 | x_1, \varnothing)$, given a report x_1. In this setup, applying the Bayesian rule, it is easy to check that $\mu(\theta_1 | x_1, \varnothing)$ is a strictly increasing function of likelihood ratio

$$\frac{\Pr(x_t = H_t | \text{Active})}{\Pr(x_t = H_t | \text{Passive})}$$

where $\Pr(x_t = H_t | \text{Active}) = \pi$,   and $\Pr(s_t = H_t | \text{Passive}) = \pi[q\sigma(H_t|H_t) + (1-q)$ $\sigma(H_t|L_t)] + (1-\pi)[(1-q)\sigma(H_t|H_t) + q\sigma(H_t|L_t)]$

First, this likelihood ratio is increasing with π and $(1-q)\sigma(H_t|H_t) + q\sigma(H_t|L_t)$. The intuition is as follows. As π increases, the probability that an active type truthfully reports H_t increases faster than that from a passive type. This is because the report from a passive type is less related to the true state compared to that from an active type. The above analysis implies, the more the investor initially believes that the true state is H_t, the more likely he will believe that the platform which reports H_t is an active type.

Second, we can differentiate the likelihood ratio with respect to the platform's strategies: $\sigma(H_t|H_t)$ and $\sigma(H_t|L_t)$. We then have the next result describing how the investor's posterior on the platform being an active type depends on the platform's reporting strategy, the accuracy of the observed signal, and the investor's priors. Summarizing the above analysis, we have the following results:

Proposition 2: At the beginning of the second period, the investor's posterior $\mu(\theta_1|x_1, \varnothing)$ has the following properties:

(1) $\mu(\theta_1|H_1, \varnothing)$ is strictly increasing with r_1, strictly decreasing with $\sigma(H_1|H_1)$ and $\sigma(H_1|L_1)$;
(2) $\mu(\theta_1|L_1, \varnothing)$ is strictly decreasing with r_1, strictly increasing with $\sigma(H_1|H_1)$ and $\sigma(H_1|L_1)$;
(3) $\mu(\theta_1|H_1, \varnothing)$ is decreasing with π and $\mu(\theta_1|L_1, \varnothing)$ is increasing with π.

If the uncertainty cannot be resolved at the end of the first period, the investor's posterior on the type of the platform would depend on the platform's reports. Then we have,

Lemma 1: In equilibrium, it is true that $\mu(\theta_1|H_1, \varnothing) = \mu(\theta_1|L_1, \varnothing)$.

4.2.2 Uncertainty Is Resolved ($\rho = 1$)
When the uncertainty is completely resolved, the passive platform faces the fact that any misreports would imply that it is a passive type. Therefore, the passive platform always prefers to truthfully report the observed signal. Intuitively, this implies that the platform on the market would always report the state which is believed to be the most possible one after observing the signal. Therefore, as long as the signal observed by the platform is strong enough, such that $q > \pi$, we would have a truthful reporting in equilibrium. We have the following result:

Lemma 2: In equilibrium, it is true that $\mu(\theta_1|H_1, H_1) = \mu(\theta_1|L_1, L_1)$.

A direct implication of this result is:

Corollary 1: If the uncertainty can be completely resolved, i.e., $\rho = 1$, and $q > \pi$, then for any type of platform, it would always truthfully report the observed signal in both periods.

5 Analysis of the General Model

In this section, we would follow backward induction to analyze the game. The solution concept we use here is Perfect Bayesian Equilibrium. For an equilibrium, the following conditions must hold: (1) in every period, each type of platform's strategy is optimal given the investor's decision rule; (2) the investor's beliefs follow the Bayesian Rule; (3) the investor's decision is optimal.

First, it is easy to check that in the second period, regardless which platform is on the market, the situation is exactly the same as the case of the monopoly market. Therefore, we have the same result as the monopoly case:

Lemma 3: In $t = 2$, for any type of platform, it would truthfully report the observed signal.

Following backward induction, we go back to the investor's decision. Given that the state of uncertainty resolution is A_i, from this point onward, we denote the representative investor's strategy after observing the report, x_t, as $d(x_t, A_i) \in \{0, 1\}$. $\mu(\theta_1 | x_t, A_i)$ is the investor's posterior on the platform being an active type after observing the report x_t in period t, where $x_t \in \{H_t, L_t\}$ is the platform's report after observing the signal ω_t, and $\mathbf{A} = \{H_1, L_1, \varnothing\}$. We have the following result:

Lemma 4: At the end of the first period, after observing the report from the platform, the representative investor's posteriors satisfy the following conditions:
$1 > \mu(\theta_1 | H_1, H_1) = \mu(\theta_1 | L_1, L_1) > \mu(\theta_1 | H_1, \varnothing) > r_1 > \mu(\theta_1 | L_1, \varnothing) > \mu(\theta_1 | H_1, L_1)$
$= \mu(\theta_1 | L_1, H_1) = 0$

The next result shows the equilibrium decisions of active platforms:

Proposition 3: Active type platforms would truthfully report the observed signal in equilibrium in both periods.

We now present our main results. In the following results, given different parameter space,

(1) for any type of platform one, there exists an equilibrium under which it would choose to truthfully report the observed signal;
(2) platform one of passive type would always report $x_1 = H_1$ if $\omega_1 = H_1$, and report $x_1 = H_1$ sometimes, if $\omega_1 = L_1$; and platform one of active type would always truthfully report the observed signal.

Formally, we have:

Proposition 4: Given $\mu(\theta_1 | L_1, \varnothing) < r_2 < \mu(\theta_1 | H_1, \varnothing)$,

(1) if $\rho > \rho^*$, there is an equilibrium under which

$$\sigma(H_1 | H_1) = 1 \text{ and } \sigma(H_1 | L_1) = 0;$$
$$d(H_1, \varnothing) = d(H_1 | H_1) = d(L_1 | L_1) = 1 \qquad \text{and}$$
$$d(H_1, L_1) = d(L_1, H_1) = d(L_1, \varnothing) = 0$$

(2) if $\rho > \rho^*$, there is an equilibrium under which
    i. when $\mu(\theta_1 | L_1, \varnothing) < r_2 < r_1$

$$\sigma(H_1 | H_1) = 1 \text{ and } \sigma(H_1 | L_1) \in (0, 1];$$

$$d(H_1, \varnothing) = d(H_1, H_1) = d(L_1, L_1) = 1,$$
$$d(H_1, L_1) = d(L_1, H_1) = 0 \text{ and } d(L_1, \varnothing) \in (0, 1)$$

    ii. when $r_1 < r_2 < \mu(\theta_1 | H_1, \varnothing)$

$$\sigma(H_1 | H_1) = 1 \text{ and } \sigma(H_1 | L_1) \in (0, 1];$$

$$d(H_1, H_1) = d(L_1, L_1) = 1$$
$$d(H_1, L_1) = d(L_1, H_1) = d(L_1, \varnothing) = 0 \quad \text{and}$$
$$d(H_1, \varnothing) \in (0, 1)$$

The above results have the following interpretations. First, given that platform one and the entrant, platform two, are similar such that $|r_1 - r_2| < \mu(\theta_1 | H_1, \varnothing) - \mu(\theta_1 | L_1, \varnothing)$, if the uncertainty is not resolved and $x_1 = L_1$, the representative investor would choose to withdraw from platform one and purchase from platform two. If the uncertainty is not resolved and $x_1 = H_1$, the representative investor would still purchase from platform one in the second period. This implies, platform two will not enter the market. Therefore, if passive platform one which believes the average quality of the risky projects is low, it would prefer not to truthfully report the signal. Because, as long as the uncertainty is not resolved, this biased report could increase the probability of keeping the investor's investment and attract more in the second period. However, this biased report could also bring costs to platform one in two ways. First, platform one would fail to get the utility from choosing the correct action, as platforms always get the largest utility by truthfully reporting the signal. Second, with probability ρ, the uncertainty is resolved, then platform one's type will be publicly revealed. The representative investor would withdraw from platform one and invest on projects on platform two. Platform one would get zero in the second period. Therefore, if ρ is small, the incentive for platform one to misreport would be high. If ρ is large, the incentive for platform one to misreport would be low.

Proposition 5: Given $r_2 > \mu(\theta_1 | H_1, \varnothing)$, there is an equilibrium under which

(1) when $\mu(\theta_1 | H_1, \varnothing) < r_2 < \mu(\theta_1 | L_1, L_1)$, we have

$$\sigma(H_1 | H_1) = 1 \text{ and } \sigma(H_1 | L_1) = 0;$$
$$d(H_1, H_1) = d(L_1, L_1) = 1 \text{ and}$$

(2) when $r_2 < \mu(\theta_1 | L_1, L_1)$, we have $d(H_1, L_1) = d(L_1, H_1) = d(H_1, \varnothing) = d(L_1, \varnothing) = 0$

$$\sigma(H_1 | H_1) = 1 \text{ and } \sigma(H_1 | L_1) = 0;$$
$$d(H_1, H_1) = d(L_1, L_1) = d(H_1, L_1) = d(L_1, H_1) = d(H_1, \varnothing) = d(L_1, \varnothing) = 0.$$

This result indicates that, when the ratio of the likelihood of being an active type between platform two and platform one is large, i.e., platform two is more likely to be an active type, we have a counter intuitive result: platform one of passive type would not misreport the observed signal even if the probability of getting uncertainty resolved is extremely small. This result is true in equilibrium, because when the investor believes that platform one is more likely than platform two to be an active type, and the uncertainty is not resolved, then the posterior, $\mu(\theta_1 | L_1, \varnothing)$, would still be larger than r_2, even if platform one misreports, i.e., $x_1 = L_1$. Thus, the representative investor will still keep the investment on platform one. In other words, platform one will be replaced by platform two only if it misreports and the uncertainty is resolved. This result shows that the competition induced by the potential entrant could lead the platform to truthfully present the state of the average quality of risky projects to investors.

Proposition 6: Given $r_2 > \mu(\theta_1|L_1, \varnothing)$, there is an equilibrium under which

$$\sigma(H_1|H_1) = 1 \text{ and } \sigma(H_1|L_1) = 0;$$

$$d(H_1, \varnothing) = d(H_1, \varnothing) = d(H_1, H_1) = d(L_1, L_1) = 1 \qquad \text{and}$$
$$d(H_1, L_1) = d(L_1, H_1) = 0$$

This result shows that, when the ratio of the likelihood of being an active type between platform two and platform one is small, i.e., platform one is more likely to be an active type, we have an opposite result. On the one hand, if the uncertainty is resolved, then the investor will not withdraw the investment only if it is revealed that platform one has truthfully reported the state of the risky projects. On the other hand, even if the uncertainty is not resolved and it misreports the state, the investor will still withdraw from platform one, because the investor's belief would be $\mu(\theta_1|H_1, \varnothing)$. Then, from Lemma 4, we know that it is less than r_2, which is the investor's belief on the entrant being an active type. Therefore, platform one's best chance of keeping the investor is to truthfully report the observed state and hope the uncertainty is resolved.

6 Policy Implications

In the above model, we argue that two factors would play key roles in determining crowdfunding platforms' information transmission, i.e., competition from entrant and ex-post uncertainty resolution. We next highlight two policy implications based on our equilibrium results.

First, in the current debate over the regulations of crowdfunding markets by the U. S. Securities and Exchange Commission (SEC), one argument is that the essence of crowdfunding is information transmission among individuals users. However, the transmission process usually comes with the reality of vulnerability to deception [17]. This argument has been proved to be true in the equilibrium of Proposition 4. Thus, the SEC's rules should pay more attention to the information transmission process in crowdfunding.

As a second implication, the effect of ex-post information variation described in Propositions 4, 5 and 6 has a direct implication to the SEC's rules which work against deception. For instance, SEC requires that crowdfunding platforms must have communication channels through which small investors can communicate with each other. This policy could help small investors obtain more information on the quality of borrowers' projects. Therefore, it could increase the probability of resolving the uncertainty. According to Propositions 4, 5 and 6, the platform's incentive of misreporting will be decreased accordingly.

7 Conclusion

In this paper, we present a new model to understand crowdfunding platforms' information reporting strategy. We find that full disclosure (truthfully reporting) may not happen in equilibrium even if we have market competition. This result does not arise from small investors' preferences on disclosed information, or small firms' ability to capture crowdfunding platforms. Instead, it arises as a result of rational platforms' desire to stay on the market for long-term returns. The advantage of our model is that it generates predictions about when biased information will arise on the market with: (1) the potential competition from new entrant, and (2) the uncertainty resolution from a third party.

Our model can be extended to include more market details. First, the information provided by platforms could be different based on their funding models (i.e., equity-based model vs. reward-based model). Thus, by further considering this difference in our model, we could generate more insight which may help us better understand this new industry. Second, crowdfunding is a product of financial innovation. The outcome of innovation strongly depends on the government's preferences and actions. Therefore, we could also consider more government behavior in the model. Overall, our model can be the basic building blocks for all these extensions.

References

1. Brandenburger, A., Polak, B.: When managers cover their posteriors: making the decisions the market wants to see. RAND J. Econ. **27**(3), 523–541 (1996)
2. Burtch, G., Ghose, A., Wattal, S.: An empirical examination of the antecedents and consequences of contribution patterns in crowd-funded markets. Inf. Syst. Res. **24**(3), 499–519 (2013)
3. Burtch, G., Ghose, A., Wattal, S.: Cultural differences and geography as determinants of online pro-social lending. MIS Q. **38**(3), 773–794 (2014)
4. Burtch, G., Ghose, A., Wattal, S.: The hidden cost of accommodating crowdfunder privacy preferences: a randomized field experiment. Manag. Sci. **61**(5), 949–962 (2015)
5. Duarte, J., Siegel, S., Young, L.: Trust and credit: the role of appearance in peer-to-peer lending. Rev. Financ. Stud. **25**(8), 2455–2484 (2012)
6. Emmerson, L.: Crowdfunding industry overtakes venture capital and angel investing (2015). http://blog.symbid.com/2015/trends/crowdfunding-industry-overtakes-venture-capital-and-angel-investing/
7. Gorton, G.B., Pennacchi, G.G.: Banks and loan sales marketing nonmarketable assets. J. Monet. Econ. **35**(3), 389–411 (1995)
8. Hildebrand, T., Puri, M., Rocholl, J.: Adverse incentives in crowdfunding. Manag. Sci. **63**, 587–608 (2016)
9. Holmstrom, B., Tirole, J.: Financial intermediation, loanable funds, and the real sector. Q. J. Econ. **112**(3), 663–691 (1997)
10. Iyer, R., Khwaja, A.I., Luttmer, E.F., Shue, K.: Screening peers softly: inferring the quality of small borrowers. Manag. Sci. **62**(6), 1554–1577 (2015)

11. Lin, M., Prabhala, N.R., Viswanathan, S.: Judging borrowers by the company they keep: friendship networks and information asymmetry in online peer-to-peer lending. Manag. Sci. **59**(1), 17–35 (2013)
12. Lin, M., Viswanathan, S.: Home bias in online investments: an empirical study of an online crowdfunding market. Manag. Sci. **62**(5), 1393–1414 (2015)
13. Liu, D., Brass, D.J., Lu, Y., Chen, D.: Friendships in online peer-to-peer lending: pipes, prisms, and relational herding. MIS Q. **39**(3), 729–742 (2015)
14. Michels, J.: Do unverifiable disclosures matter? Evidence from peer-to-peer lending. Acc. Rev. **87**(4), 1385–1413 (2012)
15. Mollick, E., Nanda, R.: Wisdom or madness? Comparing crowds with expert evaluation in funding the arts. Manag. Sci. **62**(6), 1533–1553 (2015)
16. Prendergast, C.: A theory of "yes men". Am. Econ. Rev. **83**(4), 757–770 (1993)
17. Shiller, R.J.: Opinion: do crowdfunding rules ignore human nature? (2015). http://www.marketwatch.com/story/crowdfunding-or-crowdphishing-2015-11-18

Privacy-Preserving Access Control Scheme for Outsourced Data in Cloud

Ning Zhang and Jianming Zhu[(⊠)]

School of Information, Central University of Finance and Economics,
Beijing 100081, China
zhangning75@sina.com, zjm@cufe.edu.cn

Abstract. Considering of economy, efficiency and security, more and more small and medium economic and social organizations are trying to outsource their growing business data to one or multiple professional storage service providers. But the separation between actual data owners and operator leads the demands for the technology of secure distributed storage. In this condition, we don't assume that the entity enforcing access control policies is also the owner of data and resources like in traditional access control models. In this paper, we discuss the current state of information security outsourcing and analyze outsourced data security in cloud, access control and secret sharing. And then a new privacy enhanced access control scheme for outsourced data in cloud based on secret share is presented. In this scheme, a sensitive outsourced data can be divided to many shares which are stored in multiple cloud servers separately. Furthermore, the proposed scheme uses several service providers to guarantee the availability of the services. The evaluations demonstrate that our data outsourcing framework is scalable and practical.

Keywords: Outsourcing · Access control · Information security · Secret share

1 Introduction

With the development of electronic commerce and electronic government, computing becomes more pervasive and government and businesses increasingly depend on the various information systems (IS). The IS departments of firms are also facing the issues of the cost-cutting measures that organizations are adopting to remain competitive [9]. In order to control costs and at the same time to develop their information technology (IT) and IS capabilities, more and more small and medium economic and social organizations are now considering using outsourcing vendors. Now, outsourcing is becoming an increasingly important process in modern information system. However, with the development of the cloud computing and the increasing of the amount of data

This paper was supported by the National Natural Science Foundation of China under Grant U15092145, the National Social Science Foundation of China under Grant 13AXW010, Beijing Philosophy and Social Science Foundation of China under Grant 14JGA001, Discipline Construction Foundation of Central University of Finance and Economics.

M. Fan et al. (Eds.): WeB 2016, LNBIP 296, pp. 215–224, 2017.
https://doi.org/10.1007/978-3-319-69644-7_22

that various information systems generate, outsourcing introduces new security vulnerabilities due to the corporation's limited knowledge and control of external providers operating in foreign countries [1–3, 6, 7, 10]. As firms work with cloud computing service providers, data security, privacy, and interoperability issues will also surface [6]. A crucial problem for owners is how to secure sensitive information accessed by legitimate users only using the trusted services.

Cloud computing can provide an IT infrastructure, including storage, platform, service, software, etc., and it is possible for other entities to shift their IT department to a cloud service provider [4, 5]. Data outsourcing, as one of the main components in cloud computing to share resources, provides data to customers. With the outsourcing of data to the cloud, the control of the data is also shifted to the cloud. Then security and privacy issues such as data confidentiality, data availability, data backup/recovery, data authentication, computation verifiability follows. Indeed, there are various reports of security breaches in real world usage of cloud computing [8]. All the giant cloud service providers have experienced some kind of data loss [9]. Some users data on the cloud is also reported to be leaked from time to time [10]. These incidents pose a challenge to the long-term development of cloud computing. Therefore, it is of critical importance to solve the security challenges for cloud service providers, which can attract the long-term interest of users to employ cloud computing and enable the sustainable development of cloud computing.

As one of the powerful and generalized approaches to security management, access control is used to ensure that only authorized users were given access to certain data or resources. Access control includes authentication, authorization and audit. The existing access control models tend to fall into three classes: Discretionary Access Control (DAC), Mandatory Access Control (MAC) and Role Based Access Control (RBAC).

While there are a broad range of security issues, we focus on the access control problems with regard to data and computation outsourcing in this paper. And a new privacy enhanced RBAC scheme for outsourced data in cloud is proposed.

The rest of the paper is organized as follows: we investigate the related work on the security of outsourced data in cloud in Sect. 2. In Sect. 3, we propose a new privacy enhanced access control scheme for outsourced data in cloud based on secret sharing. We give our conclusion in Sect. 4.

2   Related Work

This section presents the basic technology of outsourced data security in cloud and introduces the status of access control scheme and secret share.

2.1   Outsourced Data Security in Cloud

In 2011, Cloud Security Alliance (cloudsecurityalliance.org/guidance/csaguide.v3.0.pdf) published a document about secure data storage and management, which is an important component of cloud computing [1]. In the guidance, a secure outsourced data system should be assessed from at least the following aspects: (1) strong encryption and scalable key management; (2) de-provisioning, and information lifecycle management; and

(3) system availability and performance. The most convincing and naive solutions for these cases are to encrypt data by the owner before sending them to the server for storage and to provide the decryption key to users authorized to access the data (Fig. 1) [1].

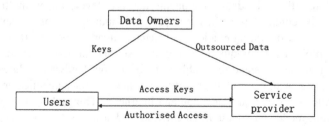

Fig. 1. Cloud computing model

Despite of the rise of cloud computing services and its widespread utility, preserving security and privacy is the main challenge in this area. Outsourced data for multi-user accesses can be achieved through encrypted file systems. However, the existing approaches show the shortages for application [1]. By outsourcing data to the cloud, users have no idea where their own data is exactly located and they no longer have physical access and full control over it.

According to cloud users, security issues are [15]:

- Due to the general lack of transparency and long term data storage in the cloud systems, how is a cloud server entirely monitored and traced, aiming to prevent wicked insider from access and misuse users' private data?
- Due to the huge amount of data stored in the cloud, how can users audit all the activities and processes done by CSPs (Cloud Service Providers) on their data in order to have more secured and trusty cloud?
- How to ensure that firm and proper laws are established in the cloud in order to protect cloud users against their cloud providers in case of any malicious and illegal data utilization by them?
- Who exactly provides and manages the encryption/decryption keys, which should be logically done by the cloud users?

To address the security and privacy issues listed above, several researchers have proposed multiple methods. For secure and efficient access to outsourced data, both data and metadata must be properly protected from unauthorized users. Wang et al. propose a new RBAC model with encryption technology for outsourced data [1]. Li et al. propose MAACS (Multi-Authority Access Control System), a novel multi-authority attribute-based data access control system for cloud storage [3]. In [7], Lei et al. are motivated to design a protocol to enable secure, robust cheating resistant, and efficient outsourcing of matrix multiplication computation (MMC) to a malicious cloud in this paper. The main idea to protect the privacy is employing some transformations on the original MMC problem to get an encrypted MMC problem which is sent to the cloud; and then transforming the result returned from the cloud to get the correct result to the original MMC problem. Sujithra et al. describe how securely the mobile data can be stored in the remote cloud using cryptographic techniques with minimal performance degradation [8]. Hamlen et al. discusses security issues pertaining to data-as-a-service and software-as-a-service models as well as supply chain security issues and present relevant standards for data outsourcing. Its goal is for the composite system to be secure even if the individual components that are developed by multiple organizations might be compromised. In paper [11], the authors propose a new data outsourcing framework providing efficient and scalable query response times. In addition to this, the proposed technique uses multiple service providers to guarantee the availability of the services and to be able to recover from hardware failures. Liu et al. bring forth a big picture through providing an analysis on authenticator-based data integrity verification techniques on cloud and Internet of Things data [14]. Sareen et al. propose a new model based on fragmentation and a secret sharing scheme which partition data among multiple cloud service providers [16].

All these approaches adopt asymmetric encryption methods to protect data from unauthorised access and with a huge key numbers. In this paper, a new privacy enhanced access control scheme for outsourced data in cloud based on secret share is presented.

2.2   Access Control

The primary goal of information security is to protect the fundamental data that powers our systems and applications. As companies transition to cloud computing, the tradition methods of securing data are challenged by cloud-based architectures. There are some of more useful technologies and best practices for securing data within various cloud models. They are content discovery, IaaS encryption, PaaS encryption, SaaS encryption.

Three primary rules are defined for RBAC (Fig. 2):

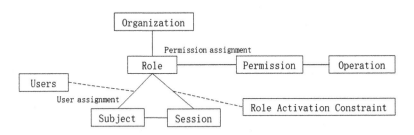

Fig. 2. Role-based access control

Rule 1. Role assignment: A subject can exercise a permission only if the subject has selected or been assigned a role.

Rule 2. Role authorization: A subject's active role must be authorized for the subject. With rule 1 above, this rule ensures that users can take on only roles for which they are authorized.

Rule 3. Permission authorization: A subject can exercise a permission only if the permission is authorized for the subject's active role. With rules 1 and 2, this rule ensures that users can exercise only permissions for which they are authorized.

In RBAC, roles are created for various job functions, and the permissions to perform certain operations are assigned to specific roles. Members of staff (or other system users) are assigned particular roles, and through those role assignments acquire the permissions to perform particular system functions.

2.3 Secret Sharing

Secret sharing refers to methods for distributing a secret amongst a group of participants, each of whom is allocated a share of the secret. The secret can be reconstructed only when a sufficient number, of possibly different types, of shares are combined together; individual shares are of no use on their own. Counting on all participants to combine the secret might be impractical, and therefore sometimes the threshold scheme is used where any k of the parts are sufficient to reconstruct the original secret.

In one type of secret sharing scheme there is one dealer and n players. The dealer gives a share of the secret to the players, but only when specific conditions are fulfilled will the players be able to reconstruct the secret from their shares. The dealer accomplishes this by giving each player a share in such a way that any group of t (for threshold) or more players can together reconstruct the secret but no group of fewer than t players can. Such a system is called a (t, n)-threshold scheme (sometimes it is written as an (n, t)-threshold scheme).

In Fig. 3, the goal is to divide secret S (e.g., a safe combination) into n pieces of data $S_1, \cdots, S_n$ in such a way that:

(1) Knowledge of any k or more S_i pieces makes S easily computable.
(2) Knowledge of any $k - 1$ or fewer S_i pieces leaves S completely undetermined (in the sense that all its possible values are equally likely).

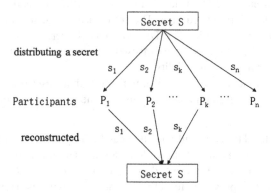

Fig. 3. Secret sharing scheme

This scheme is called (k, n) threshold scheme. If $k = n$ then all participants are required to reconstruct the secret. In this paper, we propose a new model based on secret sharing scheme which partition data among multiple cloud service providers.

3 A New Privacy-Preserving Access Control Scheme for Outsourced Data in Cloud

In this section, we define the problem and introduce the model and then we propose a new privacy enhanced access control scheme for outsourced data in cloud based secret sharing. The basic idea behind this scheme is to integrate RBAC and encryption of the outsourced data and secret sharing.

3.1 Security Model

Our model is based on distributed architecture in which a data owner wants to outsource its data $D(a_1, a_2, \cdots, a_n)$ on multiple clouds managed by different DSPs, where $a_1, a_2, \cdots, a_n$ are the attributes of data D. Some of attributes is public, some attributes are is private data and is can be accessed by authorized user, some attributes are secret data and must be accessed if most users arrive at agreement.

In our model, data reside on multiple external servers owned by different DSPs and the user access data through the Internet by an application interface.

Figure 4 describes the security model of our scheme, which is composed of the following 5 parties:

(1) Cloud Storage Server: It is an entity that stores the encrypted data and responds to access requests. Data storage server may be provided by different data service providers (DSPs).

(2) Data Owner: It creates data to be stored at remote storage in an encrypted format and regulates who has what access to each data. It can also call for the cloud server to delete the data file.

(3) Users: Each user with a Uid is labeled by a set of attributes and may have read and write access to the protected data. The set of users is denoted as U.

(4) Trusted Proxy Server (TPS): The Trusted Proxy Server acts as a supervisor in this architecture and is assumed to be trustworthy. It takes data and security parameters in the form of confidentiality.

(5) Roles: A job function within the organization that describes the authority and responsibility conferred on a user assigned to the role. The set of roles is denoted as R.

In this model, the TPS is the only one that can be fully trusted. And as similar as the assumption in [2, 3], we assume that the cloud storage server is honest but curious. The cloud server will follow the presented protocol in general, but may collude with malicious users or data providers to get illegal access privileges. However, it will not collude with the revoked users. We assume that the cloud server mostly focuses on information of data contents. We assume that the users are malicious all the time. They

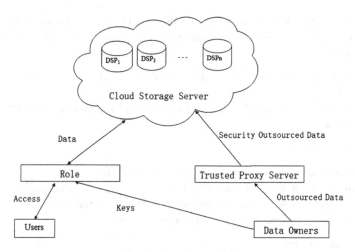

Fig. 4. Security model

may collude with the others and even the cloud server, and try to access the data that they are not authorized.

In this paper, we assume that one user receives only one key to decrypt the outsourced data stored at cloud services and one outsourced data are encrypted by one key only. There is no restriction on which encryption algorithm are applied in this paper, instead of, as a complementary, any cryptographic algorithms are applicable. Also, we borrow the existing key derivation method proposed by [1] on the basis of the key initialization, and computation of public tokens that allow the key derivation is possible from other keys.

3.2 A New Access Control Scheme for Outsourced Data in Cloud Based on Secret Sharing

In this section, a new access control scheme for outsourced data in cloud based on secret sharing is proposed.

Definition 1. The authorization models are composed of the following components:

(1) A set R of roles, a set P of permissions with a set of D as objects and OP for operations, a set K of keys and T of tokens.
(2) A set of data access permissions and permission-role assignments

$$P = \{(op, d) | op \in OP, d \in D\}$$

$$PR = \{(r, p) | r \in R, p \in P\}$$

(3) A set of data access with keys and tokens DAK = $\{(a, op, k) | d \in D, op \in OP, k \in K, t \in T\}$

(4) Key-based data access assignment KDS⊆R × DAK is a many-to-many relation that decides what roles with decryption keys can access the data based on authorizations.

Let R be the set of roles of the system and P the set of permissions on outsourced data with operations,

Now, we present the detailed of the access control scheme for outsourced data in cloud as follows.

(1) *System initialization*

The system parameters are set up by the following algorithms.

Data owner want to outsource data $D(a_1, a_2, \cdots, a_n)$ to store on cloud storage server, where $a_1, a_2, \cdots, a_n$ are the attributes of data D. Some attributes are sensitive data and they must be encrypted by data owner.

(2) *Data outsourcing*

When data owner wants to outsource it data $D(K, a_1, a_2, \cdots, a_n)$ on multiple clouds managed by different DSPs, he encrypt D using key K.

The TPS acts as a supervisor in our scheme and is assumed to be trustworthy. It takes data and security parameters in the form of confidentiality constrains from the data owner. So, it fragments the data D and stores the fragments at multiple servers in clouds.

In this scheme, the sensitive attributes ds with Constraints is protected by applying threshold secret sharing scheme. Data owner distribute a secret ds into n pieces {s1, s2, ..., sn} and store them among n data service providers {DSP1, DSP2, ..., DSPn}, such that knowledge of any t (t ≤ n) service providers is required to reconstruct the secret in addition to some secret information, X, known only by data owner.

Since even complete knowledge of t − 1 peers cannot reveal any information about the secret even though they know secret information X. This scheme is called (t, n) threshold scheme to hide the secret.

In our scheme, TPS choose a random polynomial f(x) of degree n − 1 where the constant term is the secret ds, and secret information X which is a set of n random points. Then, data owner computes the share of each service provider as f(xi) and sends it to data service provider DSPi.

Since service providers do not know each other and secret information X, they cannot find out the secret values (even if they combine their shares).

For example, there are three DSPs, i.e. DSP_1, DSP_2, DSP_3 and we use $(2, 3)$ secret share scheme, t = 2, n = 3.

– Data Owner
 • k is the key between Data Owner and TPS, $Enc_k(D)$ is encrypted data D with key k.
 • Send $Enc_k(D(d_1, d_2, \ldots, d_{s-1}, d_{s+1}, \ldots, d_n))$ to TPS
 • Send $Enc_k(d_s)$ to TPS

We assume that k_1 is key that used to encrypt and decrypt the data. Note: we borrow the existing key derivation method proposed by [1] on the basis of the key initialization, and computation of public tokens that allow the key derivation is possible from other keys.

- TPS →Cloud Storage Server
  - Send $Enc_{k1}(D(d_1, d_2, \ldots, d_{s-1}, d_{s+1}, \ldots, d_n))$ to DPS_1
  - creates a random polynomial $f(x) = a_2 x^2 + a_1 x + a_0$ with degree 2 and a constant term $a_0 = d_s$;
  - chooses secret information $X(x_1, x_2, x_3)$ which is 1 random points, such that $x_i \neq 0$;
  - Computes shares s_1, s_2, s_3 of secret d_s for DSP_1, DSP_2, DSP_3, share $s_i = f(x_i)$
  - Send $Enc_{k1}(s_1)$ to DSP_1, $Enc_{k1}(s_2)$ to DSP_2, $Enc_{k1}(s_3)$ to DSP_3
- User
  - When user want to visit the data D, Data Owner will assigns the key and token according to its role.
  - The user can compute the key k_1 and he can get the data D from the Cloud Storage Server.
  - When the user get the $Enc_{k1}(s_1)$ from DSP_1, $Enc_{k1}(s_2)$ from DSP_2, $Enc_{k1}(s_3)$ from DSP_3 respectively. And then he can decrypt them using the key k_1 and get the s_1, s_2, s_3.
  - Using k points from the set X and the values of $f(x_i)$, construct 2 points $(x_i, f(x_i))$ where $i = 1, 2$.
  - Compute the secret d_s using Newtons' divided difference interpolation and then evaluate $d_s = a_0$, which is the original value of the sensitive attribute

4 Conclusion

Protecting outsourced data stored in the cloud environments is vital, especially when it comes to confidential information. In recent years, many researchers have focus this problem and proposed various approaches to protect the security of outsourced data. However, due to the complexity of this issue, this problem has not been solved so far.

In this paper, we propose a new a new access control scheme for outsourced data in cloud based on secret share is presented. An example of how to apply this scheme to a practical system is described. The results demonstrate practical performance for many real-world applications, and this scheme is security and highly scalable.

References

1. Wang, H., Yi, X., Bertino, E., Sun, L.: Protecting outsourced data in cloud computing through access management. Concurr. Comput. Pract. Exp. wileyonlinelibrary.com. doi:10.1002/cpe.3286
2. Li, M., Yu, S., Zheng, Y., Ren, K., Lou, W.: Scalable and secure sharing of personal health records in cloud computing using attribute based encryption. IEEE Trans. Parallel Distrib. Syst. **24**, 131–143 (2013)

3. Li, Q., Ma, J., Li, R., Liu, X., Xiong, J.: Secure, efficient and revocable multiauthority access control system in cloud storage. Comput. Secur. **59**, 45–59 (2016)
4. Khalfan, A.M.: Information security considerations in IS/IT outsourcing projects: a descriptive case study of two sectors. Int. J. Inf. Manag. **24**, 29–42 (2004)
5. Bachlechner, D., Thalmann, S., Maier, R.: Security and compliance challenges in complex IT outsourcing arrangements: a multi-stakeholder perspective. Comput. Secur. **40**, 38–59 (2014)
6. Ali, S., Khan, S.U.: Software outsourcing partnership model: an evaluation framework for vendor organisations. J. Syst. Softw. **117**, 402–425 (2016)
7. Lei, X., Liao, X., Huang, T., Heriniaina, F.: Achieving security, robust cheating resistance, and high-efficiency for outsourcing large matrix multiplication computation to a malicious cloud. Inf. Sci. **280**, 205–217 (2014)
8. Sujithra, M., Padmavathi, G., Narayanan, S.: Mobile device data security: a cryptographic approach by outsourcing mobile data to cloud. Procedia Comput. Sci. **47**, 480–485 (2015)
9. Liang, H., Wang, J.-J., Xue, Y., Cui, X.: IT outsourcing research from 1992 to 2013: a literature review based on main path analysis. Inf. Manag. **53**(2), 227–251 (2016)
10. Hamlen, K.W., Thuraisingham, B.: Data security services, solutions and standards for outsourcing. Comput. Stand. Interfaces **35**(1), 1–5 (2013)
11. Emekci, F., Methwally, A., Agrawal, D., El Abbadi, A.: Dividing secrets to secure data outsourcing. Inf. Sci. **263**, 198–210 (2014)
12. Cullinan, C.P., Zheng, X.: Outsourcing accounting information systems: evidence from closed-end mutual fund families. Int. J. Acc. Inf. Syst. **17**, 65–83 (2015)
13. Martinsons, M.G.: Outsourcing information systems: a strategic partnership with risks. Long Range Plan. **26**(3), 18–25 (1993)
14. Liu, C., Yang, C., Zhang, X., Chen, J.: External integrity verification for outsourced big data in cloud and IoT: a big picture. Future Gener. Comput. Syst. **49**, 58–67 (2015)
15. Torabian, H.: Protecting sensitive data using differential privacy and role-based access control, Ph.D. thesis, University Laval (2016)
16. Sareen, S., Sood, S.K., Gupta, S.K.: Towards the design of a secure data outsourcing using fragmentation and secret sharing scheme. Inf. Secur. J. A Glob. Perspect. **25**(1–3), 39–53 (2016)

Continuous User Authentication on Touch-Screen Mobile Phones: Toward More Secure and Usable M-Commerce

Dongsong Zhang[1]([✉]), Yin Kang[1], Lina Zhou[1], and Jianwei Lai[2]

[1] Department of Information Systems, University of Maryland, Baltimore, USA
{zhangd, ky6, zhoul}@umbc.edu
[2] School of Information Technology, Illinois State University, Normal, USA
jlail2@ilstu.edu

Abstract. Recent advances in sensing and wireless communication technologies have led to an explosion in the use of touch-screen mobile devices such as smartphones and tablets in mobile commerce and other daily work and life activities. These activities have resulted in more and more private and sensitive information stored on those devices. Therefore, improving the security of mobile devices by effective user authentication to prevent unauthorized information access becomes an imminent task. Mobile user authentication refers to the process of checking a user's identity and verifying whether he/she is authorized to access a device. Due to the increasing incidence of mobile phones getting lost, stolen, or snatched while being used by the owner, continuous user authentication (CUA) after logging in a mobile device has attracted increasing attention. Prior research has shown that traditional password authentication is insufficient or ineffective for CUA. Despite the recent research progress in CUA, many existing methods are explicit by nature in that they require users to perform specific operations, which can cause interruptions to users' ongoing activities or may be easily learned from observation by others. In this research, we propose a new touch dynamics based approach to CUA on touch screen mobile devices that authenticates users while they are interacting with mobile devices. Touch dynamics, which is rich in cognitive quality and unique to individuals, has yet to be explored for implicit CUA. We conducted a longitudinal study to evaluate the proposed mobile CUA approach. The results demonstrate that our method can improve the security of CUA for touch screen mobile devices. The findings have significant implications for the security and adoption of m-commerce.

Keywords: Continuous user authentication · Touch-screen mobile devices · Touch dynamics · M-commerce

1 Introduction

Mobile handheld devices, especially smartphones, have been increasingly pervasive in our daily life. Given the ubiquity and portability of mobile devices, there has been a remarkable surge in the use of those devices in business activities [1], healthcare, and work. According to Internet Retailor 2016 Mobile 500 Guide, mobile commerce

© Springer International Publishing AG 2017
M. Fan et al. (Eds.): WeB 2016, LNBIP 296, pp. 225–236, 2017.
https://doi.org/10.1007/978-3-319-69644-7_23

accounts for 30% of U.S. e-Commerce and expected to grow much faster than traditional e-Commerce. More and more consumers make online purchases, payments, stock trading, check deposit, and wire transfer, etc. through their mobile devices [1]. Workers are also increasingly mobile nowadays, with many enterprises allowing and supporting (e.g., Bring-your-own-device programs) a growing number of employees to use mobile devices to do their work at the office, at home, and while traveling.

As mobile devices and their applications continue to evolve, an increasing amount of personal, private, and important data and information are stored on those devices, ranging from personal contact information to classified work-related documents. On the other hand, such important data and information are exposed to high security risks. First, mobile devices are easily lost or stolen. In 2013, there were 4.5 million smartphones lost or stolen in U.S. A stolen phone could be hacked, in which the personal information stored in the phone could be accessed illegally. Second, a consumer report published by CNBC on April 26, 2014 revealed that only 36% of smartphone users have set a 4-digit PIN for the access to their phone. Even if a mobile device is password protected, the password could be stolen through video surveillance and shoulder surfing when people use the device in public. As a result, securing information on mobile devices through user authentication plays a critical role in preventing unauthorized misuse or abuse of information stored in those devices.

Mobile user authentication is the process of verifying a user's legitimate right to access a mobile device [2]. Authenticating users for a mobile device or application can be quite challenging. Traditional methods for authenticating users on mobile devices are based on explicit authentication mechanisms such as passwords, pin numbers, or secret patterns. Studies have shown that users often choose a simple, easily guessed password like "12345" to protect their data [3]. Verizon's 2013 Data Breach Investigations Report confirms that weak or default passwords and stolen or reused credentials are still the main source of successful data breaches. More importantly, most mobile devices simply cannot perform continuous user authentication (CUA) – continuously verifying whether a user who was originally authenticated by passwords is still the user who is accessing a mobile device.

CUA on mobile devices is important because unauthorized individuals may improperly obtain access to personal information of a user if a password is compromised or if the user does not exercise adequate vigilance after initial authentication. In the past decade, there has been increasing research on mobile user authentication and different types of approaches have been proposed. However, current authentication approaches require users to create and remember complex passwords, or to purchase special hardware or a model that is equipped with the special hardware for biometric authentication, which remains relatively expensive. In addition, gait or gesture-based approaches require users to make certain moves, which can be easily observed by others; and context-based (e.g., location and time) approaches raise several privacy concerns [4].

To address those limitations and improve CUA on mobile devices, this research proposes a novel touch-dynamics based approach to authenticating users in a continuous and transparent manner. The proposed authentication method integrates the strengths of password, gesture, and behavioral biometrics based methods. The results of a longitudinal user study demonstrate the effectiveness of the proposed approach compared with a traditional method. This research not only makes new research contributions to CUA on mobile devices, but also improves information security and privacy protection for mobile users, which will in turn increase their adoption of m-commerce.

The rest of the paper is organized as follows. We first introduce the related work on CUA. Then, we present the proposed authentication method, followed by the description of a user study. Next, we present results and discuss the major findings of this study. Finally, we conclude the paper with the last section.

2 Related Work

In recent years, there have been growing research interests in mobile user authentication. Nevertheless, password-based authentication remains the most dominant method for access control across various systems [5]. This method authorizes users based on the matching of password (e.g., a pre-arranged combination of specific keyboard characters) or password patterns chosen by users. A password commonly comes in a text form, and the text can be entered via typing, touching, and so on.

Continuous user authentication (CUA) is of paramount importance because after the initial, point-of-entry authentication, which normally verifies a password entry from a user, the user can do whatever he/she wants without having to re-authenticate or re-verify his/her identity. In Crawford and Renaud [6] 's study, 73% of participants felt that implicit authentication based on behavioral biometrics was more secure than traditional methods such as PINs and passwords, and 90% indicated that they would consider using a transparent authentication method on their mobile device.

The state-of-the-art CUA methods for mobile devices build a classification model to determine whether the current user should be authenticated or not. The key factor in the design of a CUA method is its input features. Based on the specific features used to build CUA models, existing methods can be classified into the following categories: physiological biometric based, keystroke based, gait based, gesture-based, touch-dynamics based, and behavior profiling based methods [3, 7]. Physiological biometric based authentication methods use physical biometrics for user authentication, such as finger prints, iris scan, and face, voice, or palm recognition (e.g., [8, 9]). They are touted for high recognition accuracy and uniqueness. However, physiological biometric recognition has raised privacy concerns [10] and is vulnerable to replay attack (e.g., finger residue) and spoof attack (e.g., iris copy) [11].

Behavioral profiling based methods are based on when a user should authenticate (as opposed to how) and for which mobile application. The authentication decision depends on the confidence and sensitivity levels for each application that are stated by the user to protect sensitive applications from unauthorized use [12]. For example, Shi et al. [13] built a user profile based on a user's routine, such as location, phone calls, and application usage, and assigned a positive or negative score for each user's routine. Li et al. [14] focused on mobile users' application usage, including general application usage, voice calls, and text messaging, for building user profiles. The limitation of their approach lies in its vulnerability to the likely changes of a user's application use.

With touch-dynamics based methods (e.g., [15, 16]), a user's touch gestures and finger movements on the touch screen of a mobile device while performing some basic operations are used as a data source for user authentication. Specifically, a behavioral feature vector is extracted from such recorded screen touch data, which will then be used to train a discriminative classifier for user authentication.

However, the state-of-the-art CUA methods are better characterized as explicit rather than implicit methods. They require users to perform some specific operations explicitly, which can cause interruption to the user's ongoing activities. In addition, they are not truly continuous. For example, there is a significant gap between two consecutive incidents of biometric authentication (e.g., fingerprint). This study is aimed to address these limitations of extant CUA methods by proposing a touch dynamics based method.

3   Touch Dynamics Based Mobile User Authentication

The most common type of interaction that a user may have with a mobile device is text entry. In addition, text entry is continuous, making it a great context for building CUA models. Moreover, the touch screen of modern smartphones engenders touch dynamics, which has yet to be fully explored for authenticating mobile users.

3.1   A Touch Dynamics Method

Compared with multi-finger touch, single-figure touch is easier to learn and more robust. Among the five fingers, thumb is in an advantageous position to touch the screen of a mobile device, particularly when holding the phone using the same hand. Thumb movement patterns can be viewed as a type of digital footprint, which are believed to be rich in cognitive quality and unique to individuals [15, 17]. However, such patterns have been rarely explored in behavioral biometric authentication methods [18]. A thumb stroke based text entry method could allow us to tap into the rich information source of thumb movement patterns for mobile user authentication. Although the traditional QWERTY keyboard can be operated by a thumb, it brings usability challenges. Mobile phones have tiny buttons and crowded keypads, which are difficult to press accurately with a thumb. Furthermore, when a finger touches a screen, it will cover a part of the screen underneath, causing the visual occlusion problem [19].

To address the limitations of traditional QWERTY keyboard, we selected Escape [37], a sight-free text entry method for mobile touch-screen devices. It allows the user to type letters with one hand by pressing the thumb on different areas of the screen and performing a flick gesture. While using the Escape keyboard, the screen of a mobile device is divided into four areas, with a keypad within each area. A user can tap anywhere in an area to select the character in the center, or flick toward different directions to choose characters around the center by starting the flicking gesture within a corresponding small area.

3.2   Authentication Procedure

To use Escape for CUA, a user must first enroll himself/herself by entering some fixed text using Escape. During the enrollment process, the user's thumb dynamic behavior (e.g., time and location) is automatically tracked and recorded in a system log. These data will in turn be used to learn the user's thumb stroke patterns by training models using machine learning techniques in the next step.

The learning process starts with extracting features that characterize a user's thumb movement on the touch screen of a mobile device during text entry. We adopted stroke-level features from previous studies of touch dynamic authentication systems (e.g., [16, 21]). We group the selected features for learning user thumb stroke models in this study into the following categories:

- Timing features: duration of a thumb stroke, 20/50/80 percent deviations from a stroke, 20/50/80 percent of tp (touch point)-pairwise velocity, median velocity at the last 3 tps, average velocity, 20/50/80 percent of tp-pairwise acceleration, and median acceleration at the first 5 tps;
- Spatial features: the (x, y) coordinates of the start/end tps of a thumb stroke, direct end-to-end distance of a stroke, length of a stroke trajectory, ratio of end-to-end distance to the length of a trajectory, and mid-stroke cover area;
- Movement direction features: overall stroke direction, average stroke direction, stroke average resultant length, and the largest deviation from the overall stroke direction.
- Operation features: input area (contact area of the tps from all strokes of an operation (i.e., $|X_{rightmost}-X_{leftmost}| * |Y_{highest}-Y_{lowest}|$)), number of strokes, and character entered.

We also extracted a number of character-level keystroke dynamics features from user interactions with a Qwerty keyboard in text entry to build a keystroke-dynamics based CUA as a baseline. Those keystroke features included duration based features such as key holding time and pressure based features such as maximum, minimum, and average pressure, etc.

User authentication can be treated as a binary classification problem: either authenticated or unauthenticated. Accordingly, we applied the state-of-the-art classification techniques to build authentication models by using the selected features of user thumb strokes. The following commonly used and diverse classification algorithms were deployed, which demonstrated good performance in previous authentication studies.

- A decision tree is a non-parametric supervised learning method used for classification and regression. Its goal is to create a tree-like model that predicts the value of a target variable by learning simple decision rules from the data features [22]. A decision tree is created iteratively by choosing features as decision nodes that best split data into different categories based on metrics such as information entropy.
- Naive Bayes (NB) methods are based on applying the Bayes' theorem with a "naive" assumption of independence between every pair of features [23]. We used Gaussian NB in this study.

- Support Vector Machines (SVMs) are a non-probabilistic binary classifier [24]. An SVM model represents data samples as points in a space. The samples of different categories are separated by a clear gap that is as wide as possible. New data samples are then mapped into that same space and predicted to belong to a category based on which side of the gap they fall in.

- Neural networks have a layered structure, generally consisting of at least one input layer, one output layer, and zero or multiple hidden layers [25]. Each layer contains one or multiple nodes, which are connected to all the nodes in its adjacent layer(s). Each connection has an associated weight value, which indicates the strength and polarity of that connection. By following [26], we trained and optimized an error-back propagation neural network classifier using Particle Swarm Optimization to deal with individual users' variations in their stroke patterns.

- K-Nearest Neighbors (KNN) is a type of instance-based learning or non-generalizing learning algorithm, which does not attempt to construct a general internal model. Classification is performed through a simple majority vote of the nearest neighbors of each point. In this study, KNN was used in conjunction with DTW (Dynamic Time Warping) [15, 17], which can generate higher-level features by tracing strokes and comparing distances among strokes.

- Random Forest is an ensemble classifier. Its goal is to combine the predictions of several base estimators built with a decision tree learning algorithm in order to improve their generalizability and robustness. A random forest outputs the class that is the most frequent among the output of individual decision trees. In addition, during the construction of a tree, the split is selected as the best one among a random subset of features instead of all features [27, 28].

- AdaBoost is also an ensemble method like random forest. The core principle of AdaBoost is to fit a sequence of weak learners (i.e., models that are only slightly better than random guessing) on repeatedly modified versions of data [29, 30]. The predictions from all of them are then combined through a weighted majority vote (or sum) to produce the final prediction.

The user models constructed by the above classification algorithms will be used to accept or reject a user by monitoring his/her interaction with the touch screen of a mobile device via Escape. Escape generates a trail of richer touch dynamics than keystrokes produced while using the QWERTY keyboard. Thus, we hypothesize that *CUA based on text entry using Escape will be more secure and usable than that using a standard QWERTY keyboard for touch screen mobile devices.*

4   User Experiment

We designed and conducted controlled lab experiments to evaluate our proposed method for continuous mobile user authentication. The experiment followed a 2 by 2 repeated measures design by varying the text entry method (Escape vs. QWERTY), and screen size of mobile phones (small vs. big). We selected the QWERTY soft keyboard as the baseline for comparison. It is a standard and most commonly used keyboard on mobile devices. Additionally, the text entry features using the QWERTY keyboard were drawn from previous studies [31, 32], such as holding time and maximum pressure.

4.1 Participants

Participants were recruited from a university on the east coast of the United States. They were required to have prior experience with touch-screen mobile phones. In addition, they were expected to be familiar with the QWERTY keyboard. Because the Escape keyboard was new to all participants and the QWERTY keyboard setup in individual participants' mobile phones might vary, the participants were required to go through training and a series of practice sessions to become familiarized with entering text on the same two mobile phones using both keyboards. Such a requirement posed significant challenges to participant recruitment and retention. Eventually, seven participants (3 males, 4 females) successfully completed the study. Among them, two were between 18 and 25 years old; four were between 26 and 30 years old; and one was over 30 years old. Each participant received $200 as compensation for their effort in participating in the experiment.

4.2 System Setting

In order to examine the possible impact of screen size on user performance, both keyboard applications were installed on two Android mobile phones with different screen sizes. One was a Samsung Galaxy Note II with a 5.5" screen, and the other was a Kyocera Event phone with a 3.5" screen.

In this study, by following the guideline provided by [37], we anchored Escape in the bottom-right corner of a Samsung Galaxy Note II phone without scaling, as shown in Fig. 1. For a Kyocera Event phone, Escape fit the whole width of its screen.

Fig. 1. Screenshots of escape

4.3   Procedure

After signing a consent form, the participants went through a training session where they learned how to use each text entry method, particularly Escape. Upon the completion of training, the participants proceeded with the formal study. During this study, each participant completed tasks over 10 sessions. During each session, the participants were asked to enter short phrases, which were presented on a monitor in front of them, into a mobile phone. The entered phrases varied from 16 to 43 characters (mean = 28.61). Those phrases were randomly grouped into sets of ten. Each participant first entered one set of phrases using one of the two authentication methods and one of the two mobile phones. Then, she/he repeated the text entry process once for each of the remaining combinations of text entry method and screen size. Each phrase could only be entered once within the same session. The order of text entry methods, mobile phones, and phrase sets was counter-balanced to minimize possible learning effects.

To simulate a real-world situation where mobile users type while walking, with one hand occupied by something else, the participants were required to enter phrases while walking on a treadmill. Each participant held a phone and entered the phrases with his/her dominant hand only, while holding a remote controller that controlled the display of the subsequent phrase on a desktop monitor using the other hand, so that the participants could only interact with an experimental mobile phone with the thumb of the hand that held the phone. Depending on the participants' availability, any two consecutive practice sessions were scheduled at a 2–72 h interval. Additionally, a participant was not allowed to complete more than three sessions within the same day to avoid fatigue.

5   Evaluation and Results

5.1   Evaluation Metrics

We adopted the most commonly used metrics for measuring the security of user authentication systems: accuracy. Accuracy is defined as the percentage of authentication decisions that correctly validate or deny user access (see Eq. (1)).

$$Accuracy = \frac{|acceptedauthenticcases| + |rejectedimpostercases|}{|authenticcases| + |impostercases|} \tag{1}$$

The accuracy was reported as the average of 50 runs of 10-fold cross-validations using the data collected during the formal study.

Usability was assessed by subjective measures, specifically satisfaction and intention to use, which were collected through a post-study questionnaire adapted from [33]. The questionnaire items were rated on a seven-point Likert scale, with 1 indicating 'strongly disagree', 4 'neutral', and 7 'strongly agree'.

5.2 Results

The results of CUA are reported in Table 1. The results reveal that Escape based CUA outperforms QWERTY based method across all of the mobile authentication settings, regardless of screen size of mobile phones. Moreover, the superior performance of Escape was consistently demonstrated across all constructed classification models. Among the seven classification methods we used, random forest, neural networks, and k-nearest neighbor performed the best, and Naïve Bayes and Adaboost performed the worst, across different settings. In addition, we performed repeated measures analyses of authentication accuracy varying the authentication method and screen size. The analysis results revealed a significant main effect of authentication method ($p < .01$). Thus, our hypothesis was supported.

Table 1. Accuracy of continuous user authentication

	Escape		QWERTY	
	Big screen	Small screen	Big screen	Small screen
Decision tree	0.524	0.460	0.336	0.308
GaussionNB	0.351	0.288	0.266	0.295
Nearest neighbor	0.608	0.571	0.400	0.376
Random forest	0.604	0.560	0.376	0.360
Adaboost	0.467	0.444	0.342	0.343
SVM	0.211	0.434	0.247	0.340
RBFN	0.606	0.566	0.381	0.366
Average	0.482	0.475	0.335	0.341

A comparison of the participants' responses to the questionnaire about the two authentication methods shows that user satisfaction with Escape was lower than that with QWERTY (mean = 4.8 vs. 5.2), and their intention to use Escape was also lower than keyboard dynamics (mean = 4.0 vs. 5.1).

6 Discussion

The findings of this study demonstrate that Escape offers better security than the keystroke dynamics using a standard QWERTY keyword for mobile user authentication across device screens of different sizes. On the other hand, Escape is perceived as more difficult to use and less preferred by users.

The proposed method for mobile CUA has multiple research and practical implications.

- It addresses CUA on touch-screen mobile phones. Escape can potentially continue protecting a mobile phone even after it is lost, stolen, or snatched while being used by the owner, which has significant implications for rising mobile commerce and mobile finance.
- It improves the security of CUA on mobile devices.

- It supports one-handed mobile CUA on touch-screen mobile devices. In addition, it does not require any special hardware.
- It is transparent to users. In other words, the user is authenticated while he/she is interacting with a mobile device via text input.
- The development of user authentication models in the current study exploits a wide range of the state-of-the-art classification techniques. Our results show that, nearest neighbor, neural networks and random forest are more effective for CUA on mobile devices than alternative classification techniques.

The research findings should be interpreted in light of the following limitations of this study. It is customary to use a small sample size in longitudinal evaluations that involve multiple sessions [34]. For example, one study [34] used six and another study [35] used five participants in their longitudinal studies of a text entry method, and the third study [36] used six participants during most of the stages in testing a gesture authentication method. Nevertheless, the current study can benefit from a larger sample size, which can provide a stronger statistical power. The findings of this study echo the security-usability trade-off in the state-of-the-art mobile authentication research. Another potential reason for lower perceived usability of Escape-based CUA in comparison to that of Qwerty-based CUA is because the participants were more familiar with the latter. It would be interesting to investigate the potential learning effect of the former method over a longer period of time.

7 Conclusion

Validating a user's identity is one of the fundamental security requirements in mobile commerce and mobile banking. Mobile authentication is necessary for preventing unauthorized access to mobile devices with increasingly more personal information such as credentials and financial information stored on these devices. We proposed and evaluated a continuous mobile user authentication method based on users' single-handed interactions with touch-screen mobile phones in text entry in this study. The results of our longitudinal study show that using touch dynamics extracted from users' interaction with Escape improves the accuracy of CUA in comparison to using keystroke dynamics extracted from users' interaction with the Qwerty keyboard. Future research should investigate how to improve the usability of the text entry method to address the security-usability trade-off.

Acknowledgements. This research was supported in part by the National Science Foundation (SES-152768, IIS-1250395, CNS 1704800). Any opinions, findings or recommendations expressed here are those of the authors and are not necessarily those of the sponsor of this research.

References

1. Bhatti, T.: Exploring factors influencing the adoption of mobile commerce. J. Int. Bank. Commer. **12**, 1–13 (2007)

2. Abdulhakim, A., Abdul, M.: Touch gesture authentication framework for touch screen mobile devices. J. Theor. Appl. Inf. Technol. **62**, 493–498 (2014)
3. Patel, V.M., Chellappa, R., Chandra, D., Barbello, B.: Continuous user authentication on mobile devices: recent progress and remaining challenges. IEEE Sig. Process. Mag. **33**, 49–61 (2016)
4. Preuveneers, D., Joosen, W.: SmartAuth: dynamic context fingerprinting for continuous user authentication. In: Proceedings of the 30th Annual ACM Symposium on Applied Computing, pp. 2185–2191. ACM, Salamanca, Spain (2015)
5. Karnan, M., Akila, M.: Identity authentication based on keystroke dynamics using genetic algorithm and particle swarm optimization. In: 2nd IEEE International Conference on Computer Science and Information Technology, ICCSIT 2009, pp. 203–207 (2009)
6. Crawford, H., Renaud, K.: Understanding user perceptions of transparent authentication on a mobile device. J. Trust Manag. **1**, 1–28 (2014)
7. Al-Rubaie, M., Chang, J.M.: Reconstruction attacks against mobile-based continuous authentication systems in the cloud. IEEE Trans. Inf. Forensics Secur. **11**, 2648–2663 (2016)
8. Hadid, A., Heikkila, J.Y., Silven, O., Pietikainen, M.: Face and eye detection for person authentication in mobile phones. In: 2007 First ACM/IEEE International Conference on Distributed Smart Cameras, pp. 101–108 (2007)
9. Kim, D.J., Chung, K.W., Hong, K.S.: Person authentication using face, teeth and voice modalities for mobile device security. IEEE Trans. Consum. Electron. **56**, 2678–2685 (2010)
10. Prabhakar, S., Pankanti, S., Jain, A.K.: Biometric recognition: security and privacy concerns. IEEE Secur. Priv. **1**, 33–42 (2003)
11. Qinghan, X.: Security issues in biometric authentication. In: Proceedings from the Sixth Annual IEEE SMC Information Assurance Workshop, pp. 8–13 (2005)
12. Riva, O., Qin, C., Strauss, K., Lymberopoulos, D.: Progressive authentication: deciding when to authenticate on mobile phones. In: Proceedings of the 21st USENIX Conference on Security Symposium, p. 15. USENIX Association, Bellevue, WA (2012)
13. Shi, E., Niu, Y., Jakobsson, M., Chow, R.: Implicit authentication through learning user behavior. In: Burmester, M., Tsudik, G., Magliveras, S., Ilić, I. (eds.) ISC 2010. LNCS, vol. 6531, pp. 99–113. Springer, Heidelberg (2011). doi:10.1007/978-3-642-18178-8_9
14. Li, F., Clarke, N., Papadaki, M., Dowland, P.: Misuse detection for mobile devices using behaviour profiling. Int. J. Cyber Warf. Terror. (IJCWT) **1**, 41–53 (2011)
15. Feng, T., Liu, Z., Kwon, K.A., Shi, W., Carbunar, B., Jiang, Y., Nguyen, N.: Continuous mobile authentication using touchscreen gestures. In: 2012 IEEE Conference on Technologies for Homeland Security (HST), pp. 451–456 (2012)
16. Frank, M., Biedert, R., Ma, E., Martinovic, I., Song, D.: Touchalytics: on the applicability of touchscreen input as a behavioral biometric for continuous authentication. IEEE Trans. Inf. Forensics Secur. **8**, 136–148 (2013)
17. Feng, T., Zhao, X., Carbunar, B., Shi, W.: Continuous mobile authentication using virtual key typing biometrics. In: 12th IEEE International Conference on Trust, Security and Privacy in Computing and Communications (TrustCom). IEEE Computer Society, Los Alamitos, CA, USA; Melbourne, VIC, Australia. Country of Publication: USA. (2013)
18. Sae-Bae, N., Ahmed, K., Isbister, K., Memon, N.: Biometric-rich gestures: a novel approach to authentication on multi-touch devices. In: Proceedings of the SIGCHI Conference on Human Factors in Computing Systems, pp. 977–986. ACM (2012)
19. Scheibel, J.-B., Pierson, C., Martin, B., Godard, N., Fuccella, V., Isokoski, P.: Virtual stick in caret positioning on touch screens. In: Proceedings of the 25th IEME Conference Francophone on l'Interaction Homme-Machine, pp. 107–114. ACM, Talence, France (2013)

20. Lai, J., Zhang, D.: A study of direction's impact on single-handed thumb interaction with touch-screen mobile phones. In: CHI 2014 Extended Abstracts on Human Factors in Computing Systems, pp. 2311–2316. ACM, Toronto, Ontario, Canada (2014)

21. Trojahn, M., Ortmeier, F.: Toward mobile authentication with keystroke dynamics on mobile phones and tablets. In: 2013 Workshops of 27th International Conference on Advanced Information Networking and Applications (WAINA). IEEE Computer Society, Los Alamitos, CA, USA; Barcelona, Spain, USA (2013)

22. Mingers, J.: An empirical comparison of pruning methods for decision tree induction. Mach. Learn. **4**, 227–243 (1989)

23. Zhang, H.: The optimality of naive bayes, In: Barr, V., Markov, Z., (eds.) FLAIRS Conference, AAAI Press (2004)

24. Smola, A., Schölkopf, B.: A tutorial on support vector regression. Stat. Comput. **14**, 199–222 (2004)

25. Zhou, L., Burgoon, J.K., Twitchell, D.P., Qin, T., Nunamaker Jr., J.F.: A Comparison of classification methods for predicting deception in computer-mediated communication. J. Manage. Inf. Syst. **20**, 139–166 (2004)

26. Meng, Y., Wong, Duncan S., Schlegel, R., Kwok, L.-f.: Touch gestures based biometric authentication scheme for touchscreen mobile phones. In: Kutyłowski, M., Yung, M. (eds.) Inscrypt 2012. LNCS, vol. 7763, pp. 331–350. Springer, Heidelberg (2013). doi:10.1007/978-3-642-38519-3_21

27. Breiman, L.: Random forests. Mach. Learn. **45**, 5–32 (2001)

28. Geurts, P., Ernst, D., Wehenkel, L.: Extremely randomized trees. Mach. Learn. **63**, 3–42 (2006)

29. Freund, Y., Schapire, Robert E.: A desicion-theoretic generalization of on-line learning and an application to boosting. In: Vitányi, P. (ed.) EuroCOLT 1995. LNCS, vol. 904, pp. 23–37. Springer, Heidelberg (1995). doi:10.1007/3-540-59119-2_166

30. Zhu, J., Zou, H., Rosset, S., Hastie, T.: Multi-class adaboost. Stat. Interface **2**, 349–360 (2009)

31. Sen, S., Muralidharan, K.: Putting 'pressure'on mobile authentication. In: 2014 Seventh International Conference on Mobile Computing and Ubiquitous Networking (ICMU), pp. 56–61. IEEE (2014)

32. Hwang, S.-S., Cho, S., Park, S.: Keystroke dynamics-based authentication for mobile devices. Comput. Secur. **28**, 85–93 (2009)

33. Davis, F.D.: Perceived usefulness, perceived ease of use, and user acceptance of information technology. MIS Q. **13**, 319–340 (1989)

34. MacKenzie, I.S., Soukoreff, R.W., Helga, J.: 1 thumb, 4 buttons, 20 words per minute: design and evaluation of H4-writer. In: Proceedings of the 24th Annual ACM Symposium on User Interface Software and Technology, pp. 471–480. ACM, Santa Barbara, California, USA (2011)

35. Isokoski, P., Raisamo, R.: Device independent text input: a rationale and an example. In: Proceedings of the Working Conference on Advanced Visual Interfaces, pp. 76–83. ACM, Palermo, Italy (2000)

36. Niu, Y., Chen, H.: Gesture authentication with touch input for mobile devices. In: Prasad, R., Farkas, K., Schmidt, Andreas U., Lioy, A., Russello, G., Luccio, Flaminia L. (eds.) MobiSec 2011. LNICSSITE, vol. 94, pp. 13–24. Springer, Heidelberg (2012). doi:10.1007/978-3-642-30244-2_2

37. Banovic, N., Yatani, K., Truong, K.: Escape-keyboard: a sight-free one-handed text entry method for mobile touch-screen devices. Int. J. Mob. Hum. Comput. Interact. **5**(3), 42–61 (2013)

Erratum to: Electronic Word of Behavior: Conceptual Framework and Research Design for Analyzing the Effect of Increased Digital Observability of Consumer Behaviors in a Movie Streaming Context

Katrine Kunst[1]([✉]), Ravi Vatrapu[1,2], and Abid Hussain[1]

[1] Department of Digitalization, Centre for Business Data Analytics, Copenhagen Business School, Howitzvej 60, 4th floor, 2000 Frederiksberg, Denmark
{kk.digi, rv.digi, ah.digi}@cbs.dk,
katrine.kunst@gmail.com
[2] Westerdals - Oslo School of Arts, Communication & Technology, Oslo, Norway

Erratum to:
Chapter "Electronic Word of Behavior: Conceptual Framework and Research Design for Analyzing the Effect of Increased Digital Observability of Consumer Behaviors in a Movie Streaming Context" in: M. Fan et al. (Eds.): Internetworked World, LNBIP 296, https://doi.org/10.1007/978-3-319-69644-7_9

In the original publication of this chapter the title was incompletely shown. This has now been corrected.

The updated online version of this chapter can be found at
https://doi.org/10.1007/978-3-319-69644-7_9

Author Index

Printed in the United States
By Bookmasters